Bestiality and Zoophilia

Bestiality and Zoophilia
Sexual Relations with Animals

Edited by Andrea M. Beetz and Anthony L. Podberscek

BERG
Oxford • New York

First published in 2005 by Purdue University Press

This impression published in 2009 by
Berg
Editorial offices:
First Floor, Angel Court, 81 St Clements Street, Oxford OX4 1AW, UK
175 Fifth Avenue, New York, NY 10010, USA

Berg is the imprint of Oxford International Publishers Ltd.

Library of Congress Cataloging-in-Publication Data

A catalogue record for this book is available from the Library of Congress.

British Library Cataloguing-in-Publication Data

A catalogue record for this book is available from the British Library.

ISBN 978 1 84788 354 4 (Paper)

Printed in the UK

www.bergpublishers.com

Contents

Preface

First of all, I would like to thank the editor of *Anthrozoös*, Dr. Anthony Podberscek, for his invitation to edit this special issue on bestiality and zoophilia and thus providing a platform for the presentation and discussion of this very delicate subject, one of the last taboos.

Even though sexual contact with animals has occurred throughout history, it has rarely been investigated in much detail. Only during the last decade has it received serious attention, and studies assessing it on its own or as one amongst many behaviors have now been conducted. As editor of this special issue I consider myself lucky to have found authors who are not only experts in their fields and have conducted their own research on bestiality and zoophilia, but who are willing to contribute and share their knowledge with such a large audience. I know that sometimes very critical and negative reactions follow publications on bestiality and zoophilia and that the authors of these have sometimes been frowned upon by colleagues. However, a critical discussion is what the phenomenon of sexual contact with animals really needs; a gathering of knowledge and thoughts from different disciplines, such as psychology, mental health, law, veterinary medicine, animal protection, and sociology.

In the first article, Dr. Hani Miletski provides an extensive review of bestiality throughout history, from prehistoric times to recent developments, and in different countries and societies. The information leaves no doubt that the practice is neither new nor rare, and that it possesses many facets that need to be explored further.

The second article, by Dr. Gieri Bolliger and Dr. Antoine Goetschel, is the most comprehensive review and discussion of legislation on bestiality that has been written so far. The authors describe the bestiality laws of a number of different countries, such as Switzerland, Germany, the UK and the United States, and critically review the basis for legislation on sexual contacts between humans and animals. Special attention is given to the question of consent, and the concept of the "dignity of the animal" is introduced.

In the third article, I address the question of a possible connection between the practice of bestiality and interpersonal violence. A review of research that includes information on such a possible link is given, and an attempt to show the necessity for a more differentiated view of sexual acts with animals, that range from non-violent acts (such as fondling) to sadistic and very harmful acts, is made.

The fourth article, a reprint of the most recent research on bestiality from the veterinary sciences, by Dr. Helen Munro and Dr Michael Thrusfield, describes data on sexual abuse of pets which was obtained through a survey of UK veterinarians. While pets, most frequently dogs and cats, with severe injuries due to sexual abuse may well be seen in veterinary practices, those with minor injuries will remain undetected. The authors also point out the connection between sexual abuse of a family pet and domestic violence, in particular the abuse of children.

In the fifth contribution, Hani Miletski presents data from her own research on bestiality and zoophilia, and argues that in some cases the criteria for zoophilia as a sexual orientation are met. A sexual orientation consists of an orientation of sexual fantasy, as well as an affectional and erotic - meaning actual sexual behavior and arousal—orientation towards a certain object. Her data support the existence of such a "zoosexual orientation," as the major proportion of participants in her study reported sexual animal fantasies, sexual acts with animals, and an intense emotional involvement with some of the animals they engaged with sexually.

In the sixth article, I provide an overview of the latest research and insights into the phenomena of bestiality and zoophilia. In particular, the studies conducted by Miletski (2002), Williams and Weinberg (2004), and myself (Beetz 2002) are presented with a special emphasis on the species involved in sexual acts with humans, the sexual practices, emotional involvement with the animals, motivations, and preferences in regard to animals and humans as sex partners. Furthermore, attitudes of philosophers and society towards the practice of bestiality are discussed.

The last contribution is a commentary by Dr. Frank Ascione, who not only summarizes and critically discusses the information presented in the other articles, but also makes suggestions for further research; for example, the effects on children when their parents practice bestiality.

I would like to thank all contributing authors for their time, effort and excellent work. We now have a very useful collection of papers from which scholarly discussions on sexual contacts with animals can be generated. Comments from colleagues from different fields are always welcome, and I hope this special issue contributes to a better understanding of the complexity of bestiality and zoophilia. Finally, I would like to thank Dr. Erhard Olbrich, Dr. Hans-Georg Kluge, and Dr. Glyn Collis for refereeing the papers.

Andrea M. Beetz
Erlangen, Germany

A history of bestiality

Hani Miletski
Bethesda, Maryland, USA

Abstract

Human sexual relations with animals, a behavior known as bestiality, have existed since the dawn of human history in every place and culture in the world. Furthermore, an abundance of folklore, paintings, sculptures, films, literature and pornography exists dealing with bestiality themes. This article describes the highlights of the history of bestiality in various cultures, based on Miletski's recent book (2002).

Keywords: *Ancient Greece, bestiality, history, internet, zoophilia*

Based on the literature, bestiality—human sexual relations with animals, has been part of the human race throughout history, in every place and culture in the world. This article describes the highlights of the history of bestiality as it appears in art, folklore, religion, law, and in actual behaviors. All the facts and opinions presented in this article are taken from the literature (Miletski 2002). Most of the material reviewed and discussed is anecdotal, some is unbelievable, and occasionally authors provide conflicting data. It is important to take into consideration that some of the facts and views presented came from works that are questionable with regard to their validity. Nevertheless, the abundance of information from all around the world leaves no doubt that bestiality has been an integral part of human life.

Prehistoric Times

According to Rosenberger (1968), the practice of human–animal sex began at least in the Fourth Glacial Age, between 40,000 and 25,000 years ago. Many discoveries of paintings and carvings showing humans and animals having sexual relations have been made in various ancient religious temples (Davis 1954), indicating the pre-occupation of ancient man with bestiality (Stekel 1952). For example, according to Taylor (1996), an engraved bone rod from the cave of La Madeleine, France, from the later Ice Ages (around 25,000 years ago), depicts a lioness lick-

Address for correspondence: 6917 Arlington, Rd., Suite 202, Bethesda, Maryland 20814, USA. Ph: 301-951-6592; fax: 301-309-0446; e-mail: Hani@DrMiletski.com.

ing the opening of either a gigantic human penis or a vulva, and an Iron Age cave painting from the seventh century BC, from Val Camonica, Italy, portrays a man inserting his penis into the vagina or anus of a donkey (Gregersen 1983; Taylor 1996). According to Waine (1968), cave drawings of the Stone Age leave no doubt that our prehistoric ancestors enjoyed frequent and pleasurable sexual relations with animals. Moreover, the fact that these drawings had an integral part in a clan's family history, indicates it was a common act (Waine 1968; Ellison 1970), or at least a desired act.

Ancient Near East

Archeological findings demonstrate that bestiality was practiced in Babylonia, the ancient Empire in Mesopotamia, which prospered in the third millennium BC. In his famous Code of Hammurabi, King Hammurabi (1955–1913 BC) proclaimed death for any person engaging in bestiality (Ellison 1970; Blake 1972; Hamilton 1981). At other times, according to Waine (1968), during the Spring Fertility Rites of Babylon, dogs and other animals were used for maintaining a constant orgy condition for seven days and nights (Waine 1968).

The Hittites, (around 13th century BC), the predecessors of the Hebrews in the Holy Land, had certain rules about which animals were tolerable to have sex with and which were forbidden and punishable by death (Gregersen 1983; Kinsey, Pomeroy and Martin 1948).

The Book of Leviticus states that bestiality was very wide-spread in the country of Canaan (Dubois-Desaulle 1933; Niemoeller 1946a). The Hebrews took issue with all the previous inhabitants of the Holy Land and their customs. Even depicting God with an animal's head or an animal's body was an abomination (Dekkers 1994). The Hebrews considered sexual relations with animals a form of worshiping other Gods, as was homosexuality, and the bestialist and the animal were both to be put to death. The purpose of these taboos was to help maintain and reinforce the boundaries of the group, and enable it to retain its distinctive identity under adverse circumstances (Davies 1982).

There are four references concerning men who have sexual contacts with animals in the Old Testament (Exodus 22:19; Leviticus 18: 22–24; Leviticus 20: 15–16; Deuteronomy 27:21), and two references concerning women (Leviticus 18:23; Leviticus 20:16). The Talmud, a commentary on the Old Testament, specifically forbids a widow from keeping a pet dog, lest she be tempted to have sexual relations with it (Hunt 1974; Bullough 1976; Gregersen 1983; Dekkers 1994).

Ancient Egypt

The ancient Egyptians worshiped Gods with animal shapes almost exclusively in the pre-dynastic period before about 3000 BC (Douglas 1992). Animal–human sexual contacts are occasionally portrayed on the tombs (Bullough 1976), and bestiality was recorded in Egyptian hieroglyphics as far back as 3000 BC (Ramsis 1969). Several kings and queens had a reputation of engaging in bestiality (Rosenfeld 1967; Rosenberger 1968), most famous was Cleopatra, who was said to have had a box filled with bees which she had placed against her genitals for stimulation, similar to a vibrator (Love 1992).

Egyptian men often had sexual intercourse with cattle or any other large domesticated animal (Tannahill 1992), while the women resorted to dogs (Rosenfeld 1967; Rosenberger 1968). Sexual contacts with apes were further reported for both men and women (Masters 1962; Bagley 1968; Ramsis 1969), and most interestingly, the Egyptians are reported to have mastered the art of sexual congress with the crocodile. This was accomplished by turning the creature onto its back, rendering it incapable of resisting penetration. This form of copulation was believed to bring prosperity and restore the potency of men (Masters 1962; Bledsoe 1965; Maybury 1968; Kullinger 1969; Ramsis 1969; Love 1992). The Egyptians were also known to engage in worshipful bestiality with the Apis bull in Memphis, Egypt (*The Wild Animal Revue* 1992, Issue 5) and with goats at the Temple at Mendes (Mantegazza 1932; Bloch 1933; Niemoeller 1946a; Davis 1954; Masters 1962; Rosenfeld 1967; Bagley 1968; Maybury 1968; Love 1992). The goats were further used as a cure for nymphomaniacs (Masters 1962; Bagley 1968). Having said all that, bestiality was however, punishable in Egypt, by a variety of torture mechanisms, leading to death (Rosenfeld 1967; Rosenberger 1968; Trimble 1969).

Ancient Greece

Bestiality themes were very popular in Greek mythology (Kinsey et al. 1953; Masters 1962; Masters 1966; Rosenfeld 1967; Bagley 1968; Harris 1969; Haeberle 1978; *The Wild Animal Revue* 1991, Issue 1). Most notorious are the stories of Leda and the swan (Zeus), and Pasiphae, the wife of Minos, King of Crete, who fell in love with a bull, hid inside a wooden cow and copulated with it (Masters 1962; *The Wild Animal Revue* 1991, Issue 1; Dekkers 1994).

The worship of bulls as fertility symbols was widespread in Crete and elsewhere long before the Greek period in classical times, and the tone of the writers of the day leaves no room to doubt that bestiality was a fairly common occurrence in daily life (Niemoeller 1946a). The ancient Greeks engaged in bestiality during religious celebrations such as the Bacchanalia, in honor of the God Bacchus (Dubois-Desaulle 1933), and in the Temple of Aphrodite Parne, the Greek Goddess of Indecent Copulation (Waine 1968). As with the ancient Egyptians, ancient Greeks believed bestiality cured nymphomaniacs (Masters 1962; Bagley 1968). Bestial affairs were acted out on the Greek stage (Masters 1966), and were the theme of *The Golden Ass* by Lucius Apuleius, the earliest Latin novel that has remained in its entirety, and has long been censored because of its pornographic language and bestiality content (Ramsis 1969). Bestiality was never punishable in ancient Greece (Rosenberger 1968).

Ancient Rome

Roman mythology is rich with bestiality themes (Kinsey et al. 1953; Harris 1969), as is Greek mythology, and the Romans liked to view on stage scenes from the sexual lives of the mythological Gods, including bestial acts (Masters 1962; Bagley 1968; *The Wild Animal Revue* 1992, Issue 5).

Bestiality was wide-spread among shepherds (Masters 1962), and Roman women were known to keep snakes which they trained to coil around their thighs and slide past the lips of their vaginas (Davis 1954; Masters 1962; Christy 1967; Maybury 1968; Dekkers 1994). It was the Romans who invented the rape of women (and sometimes men) by animals for the amusement of the audience at the Coliseum and Circus Maximus, and bestiality flourished as a public spectacle in ancient Rome (Masters 1962; Bledsoe 1965; Somers 1966; Rosenfeld 1967; Rosenberger 1968; Waine 1968; Harris 1969; Trimble 1969; Blake 1972; Love 1992; *The Wild Animal Revue* 1992, Issue 5; Dekkers 1994).

Emperors, such as Tiberius (AD 14–37), his wife Julia, Claudius (AD 37–41), Nero (AD 54–68), Constantinus (a.k.a. Constantine the Great, AD 274–337), Theodora (Emperor Justinian's wife, AD 520s), and Empress Irene (AD 797–802), had been known to either engage in bestiality or enjoy watching others engage in bestiality (Bledsoe 1964; Rosenfeld 1967; Rosenberger 1968; Waine 1968; Blake 1972; Hamilton 1981; Love 1992; *The Wild Animal Revue* 1992, Issue 5).

According to Niemoeller (1946a), at the beginning of the Roman Empire, legal retribution for bestiality was required only for sodomy, under which bestiality was included. Later, bestiality was distinguished from sodomy and made punishable by death (Dubois-Desaulle 1933; Niemoeller 1946b). In any event, as the Empire expanded and grew more powerful and corrupt, punishments for bestiality became almost nonexistent (Rosenberger 1968).

The Middle Ages in Europe

Bestiality was most widespread and accepted in Western society during the Middle Ages—from the fall of the Roman Empire in AD 476 to the discovery of America by Columbus in 1492 (Dubois-Desaulle 1933; Rosenfeld 1967; Rosenberger 1968). Animals were common, everywhere, and they often shared the same roof with their owners. Sexual intercourse with animals was further thought to have been healthy and a cure for many diseases (Rosenfeld 1967; Rosenberger 1968). Nonetheless, bestiality was invariably connected with black magic and witchcraft (Niemoeller 1946a; Greenland 1983).

In the Middle Ages, bestiality received full attention from Catholic jurist-theologians, whose discussions of the matter filled volumes. One of the greatest problems involved the distinction between sexual intercourse with animals and sexual intercourse with demons, which often assumed animal form for the purpose of consorting with witches (Masters 1962). According to Salisbury's (1994) analysis of the relationship between the Church and bestiality, early Christian thinkers inherited two main traditions: (1) In the Germanic myths, heroes were described as having characteristics of strength as a result of having an animal ancestor, as was the founder of the Danish royal house who was said to have been the offspring of a bear and a woman (Liliequist 1988; Salisbury 1994). And, (2) in the classical Greco-Roman tradition, Gods appeared regularly as animals to have intercourse with humans. As the early Church fathers wrestled with this classical heritage and selected those elements suitable for Christianity, they shaped their Christian texts with the notion that humans and animals were separate, and humans should thus not have sexual relations with animals (Salisbury 1994). They made the Hebrews' laws against bestiality stricter, since bestiality did not serve reproduction (Dekkers 1994), and formal conciliar decrees began regulating sexual behavior, prescribing various penalties for bestiality (Salisbury 1994).

The early pagan Germanic secular law did not prohibit bestiality. However, as soon as Christian legislation appeared, prohibitions against bestiality emerged, suggesting that the activity was indeed going on (Salisbury 1994). The penitentials began in Ireland as a way to offer the Churchmen manuals for "healing" the souls of sinful parishioners. The early Germanic world viewed animals primarily as property and food, and this attitude was reflected in the view of the early Irish penitentials, which ranked bestiality close to masturbation, making it a mild sexual sin (Salisbury 1994). For example, an early Welsh penitential, the Preface of

St. Gildas (495–570), required a year of penance to expiate the sin of bestiality. However, if the man had been living by himself when he "sinned," three 40-day periods of fasting served as sufficient penance (Hamilton 1981). Factors of age, marital status, and ecclesiastical rank served to increase or decrease penances for all sexual sins (Salisbury 1994).

The casual attitude toward animals and sexual relations with them began to change as the conciliar legislation from the East began to influence the penitential compilers. The Council of Ancyra equated bestiality with homosexuality, and this association reached Visigothic Spain as early as the late sixth century with Martin of Braga. This shaped the Spanish penitentials from the seventh or early eighth centuries, which gave a 20-year penance for those who committed either sodomy or bestiality (Salisbury 1994). The later Irish penitentials slowly became influenced by the Council of Ancyra. Equating homosexuality with bestiality not only increased the penalty, but it communicated a change in the way people looked at animals. Instead of being an irrelevant object, the animal became a participant as in the equivalent of a homosexual encounter, and it became important to kill the animal, in order to erase any memory of the act (Salisbury 1994).

A major question, which pre-occupied the inquisitors, judges, theologians, and those who condemned witches, was whether the union of male or female witches with the Devil, under the disguise of an animal, was able to produce any offspring (Dubois-Desaulle 1933). Twelfth-century people seemed to worry more about demons than before, and tales about half-human births, which resulted from such unions, began spreading (Dubois-Desaulle 1933; Masters 1966; Rosenfeld 1967; Blake 1972; Allen 1979; Salisbury 1994; Montclair 1997).

By the 13th century, the animal world seemed much more threatening than it had been in the early Middle Ages, and penalties for bestiality increased. The idea that sexual union between man and animal may result in offspring shaped the composition of the *Summae Theologica*, by Thomas Aquinas (1224–1274), which represented the highest development of medieval thought (Salisbury 1994).

St. Thomas identified four kinds of unnatural vice: the most serious sin against nature was bestiality, followed by homosexuality, intercourse in an "unnatural position" (anything other than the missionary position), and masturbation. The attitudes of St. Thomas tended to dominate all thinking on sexual behavior to the end of the Middle Ages, resulting in classifying as deviant any kind of non-procreative sexual activity (Hamilton 1981; Salisbury 1994). Although the courts were more preoccupied with prosecuting homosexuality (Salisbury 1994), Dubois-Desaulle (1933), Niemoeller (1946a), Evans, (1987), and Dekkers (1994) describe the various types of torture accused bestialists endured, usually in the town square in front of a crowd, until they died.

The Renaissance Period in Europe

During the late Middle Ages and Renaissance, authorities were actively prosecuting homosexuals and bestialists, and the high point of bestiality trials coincided with that of the witch-hunts (Monter 1981; Dekkers 1994). During the 15th and 16th centuries, sexual relations with animals formed one of the main topics for preachers (Davies 1954; Rosenfeld 1967; Rosenberger 1968), and by 1534, bestiality became a capital crime in England and Sweden (Salisbury 1994). In 1683, Denmark passed a law making both homosexuality and bestiality punishable by burning. In 1711, it was decided that those convicted should be strangled as well as burned (L'Etalon Doux 1996). During the 17th century, the incidence of bestiality between young boys and cows and sheep became so prevalent that the Catholic Church tried to ban the employment of male herdsmen (Love 1992).

Hundreds of reports have survived from the boom in bestiality trials from the 16th to the 18th centuries (Dekkers 1994), demonstrating that bestiality was well-established in ordinary life in Europe (Niemoeller 1946a; Davis 1954). Tales about monster-looking births continued to spread (Bagley 1968; Blake 1972), as well as myths about the connection between bestiality and sexually transmitted diseases (Masters 1962; Bagley 1968). One of the persistent legends of history attributes the death of the Russian Empress Catherine the Great to an accident while attempting to have sexual relations with a bull or a horse. The sling broke, and the weight of the animal crushed her (Friday 1981; Bullough 1994; Cornog and Perper 1994).

Europe in the Modern Era and Today

Parisian brothels were known to provide turkeys for their clients. As the men were about to experience orgasm through having intercourse with the turkey, they would break the bird's neck, causing the bird's cloacal sphincter to constrict and spasm, clamping down on their penises and creating pleasurable sensations. A similar activity was enjoyed by ancient Chinese men, whose animal of preference was the goose (Edwardes and Masters 1977; Love 1992; Dekkers 1994). In 19th century France, bestiality became an organized practice, and at the time of Napoleon III, bestiality was said to have been one of the allied activities of the Society for the

Advancement of Sodomy, which met regularly in the Allees des Veuves (Dubois-Desaulle 1933; Niemoeller 1946a). According to Somers (1966), notable public displays of bestiality was one of the many factors which led to the French Revolution. In 1810, the Napoleonic code decriminalized sexual behavior done in private between consenting adults, and bestiality was no longer considered a crime (Gregersen 1983).

According to Dekkers (1994), since that time, the subject of bestiality has never been included in the civil code of France, Italy, Spain, Portugal, Romania, Belgium, and the Netherlands (Dekkers 1994), although certain acts continued to be punishable, if they involved violence or occurred in a public place. Many western countries, with the exception of a few such as England and the United States, had followed suit, at least in the elimination of the death penalty (Gregersen 1983).

In 1821, the law in England called for the death penalty for any person who committed the crime of sodomy, either with a man or with any animal. This law was revised in 1861, and the sentence reduced to life imprisonment (L'Etalon Doux 1996). Nevertheless, since in England bestiality has been lumped together with homosexuality as "sodomy;" the prosecution of the former has declined with that of the latter (Dekkers 1994). In 2002, the United Kingdom's Home Office reported the sentencing to be reduced to a maximum of two years of imprisonment (2002).

According to Dubois-Desaulle (1933), the Hungarian penal code of 1878 carries the maximum penalty of one-year imprisonment for sexual relations with animals. The German penal code of 1871, revised in 1876, in its Article 175 states that acts against nature with animals shall be punished by imprisonment, and the convicted individual shall be deprived of his civil rights (Dubois-Desaulle 1933). Bestiality stopped being a crime in West Germany in 1969 due to "lack of use" (Dekkers 1994; L'Etalon Doux 1996). In the former Eastern (communist) half of Germany, bestiality was not considered an offense (L'Etalon Doux 1996).

During World War II, human–animal breeding experiments were conducted by the Nazi physician, Dr. Josef Mengele. He was reported to be obsessed with bestiality, and was bent on creating a hybrid that could eventually replace slave labor for menial tasks. He used the large camp source of young Jewish and Polish girls in the Auschwitz concentration camp for this purpose, and forced dogs and ponies on these women. (Rosenfeld 1967; Blake 1972; *The Wild Animal Revue* 1992, Issue 7). Klaus Barbie, the infamous "Butcher of Lyons," used to force female prisoners to perform sex acts with animals as a way of degrading them, according to war crimes testimony (*The Wild Animal Revue* 1992, Issue 7).

According to Rosenberger (1968), bestiality is very common in Europe. As late as in the sixties, in Sicily and parts of France, Germany, and Poland, priests used to ask in the confessional if one had used an animal for "bestial purposes of sex" (Rosenberger 1968, p. 28). In the forties and fifties, in Sicily and southern Italy, bestiality among herdsmen was said to have been of such proportions that it was considered a national custom (Niemoeller 1946a; Davis 1954). And, Aleister Crowley, the organizer of "Love is the law" cult in Sicily, was said to have his mistress and other female members of the cult engage in acts of bestiality with his selected sacred goat (Bagley 1968).

According to a 10-year-old issue of *The Wild Animal Revue*, a specialized magazine about bestial sex, interested individuals can find sex shows involving women engaging in sexual activities with a variety of animals, such as dogs, goats, snakes, donkeys, bulls, and ponies, almost all over Europe: in Italy, Greece, Cyprus, Spain, Portugal, England, France, Germany, Austria, Norway, and in the Netherlands (*The Wild Animal Revue* 1994/95, Issue 11). Denmark is probably the only place where bestiality videos are legally produced and distributed (*The Wild Animal Revue* 1991, Issue 1), while in Hungary, magazines dedicated to animal sex are sold openly in book-stores (Byrd 2000).

South and East Asia and Oceania

In its 17 volumes—*Thousand Nights and a Night*, translated by the British explorer and orientalist, Sir Richard F. Burton, and published between the years 1885 and 1888—bestiality among Chinese with ducks, goats, and other animals, is discussed (Trimble 1969; Edwardes and Masters 1977). According to Waine (1968), in China, sexual relations with canines prospered both in the past and present. In old Shanghai, the exhibit of a young virgin being mounted by a dog was regularly offered in the brothel's sex shows (*The Wild Animal Revue* 1992, Issue 6), and Prince Chien, of the Han dynasty (221 BC–AD 24), was said to have forced women to have intercourse with dogs (Bullough 1976). Sultans and other leaders of the East were said to use animals to keep the women of their harems happy and satisfied (Somers 1966; Bagley 1968; Trimble 1969). In ancient days, Pekingese dogs were bred and raised by eunuchs under close supervision of the Emperor himself. The royal preference for Pekingese probably precluded penetration possibilities, but the special treatment given to their tongues by the eunuchs, and the common practice of puppy breast-feeding by privileged ladies, indicate dog–human sexual attitudes "beyond the shadow of a doubt" (Waine 1968, p. 49). The Pekingese was replaced by the Chow Chow as Imperial Dog in following centuries (Waine 1968).

As mentioned before, the wealthy and sophisticated men of the East, especially the Chinese, were famous for their intimate relations with geese and other birds, whose necks they wrung at the moment of orgasm in order to obtain added stimulation from the final spasms of the animal's anal sphincter (Mantegazza 1932; Davis 1954; Masters 1962; Greenwood 1963; Rosenfeld 1967; Rosenberger 1968; Ramsis 1969; Mantegazza 1886 cited in Edwardes and Masters 1977; Dekkers 1994).

In 1933, Dubois-Desaulle stated that bestiality was popular in the "Orient." Before communism in China, almost any sex show could be seen in Shanghai. Yet, currently no animal sex shows are known to take place in China (*The Wild Animal Revue* 1994/95, Issue 11). However, *The Wild Animal Revue* (1994/95, Issue 11) relates that in Southeast Asia one can find sex shows with barnyard, domestic animals, snakes, and eels. Thailand is notorious for its human–animals sex shows, as are Singapore, Taiwan, Korea, the Philippines, and the former French colonies of the Pacific Islands (*The Wild Animal Revue* 1994/95, Issue 11).

Although there is very little bestiality among the Japanese (Rosenberger, 1968), the ultimate bachelor party extravaganza in modern day Japan is said to be an exhibit of a young woman being mounted by a dog, and underground bestial sex shows can still be found (Trimble 1969; *The Wild Animal Revue* 1992, Issue 6; *The Wild Animal Revue* 1993, Issue 9).

In Australia and New Zealand, dog, goat, ram, pony, and bull sex shows exist (*The Wild Animal Revue* 1994/95, Issue 11). It is also reported that the Aborigines of Australia are known to practice bestiality (Rosenberger 1968).

Bestiality was very common among the Hindus (Rosenfeld 1967), and portrayals of animal–human sexual contacts frequently appear in temple sculptures all over India (Rosenfeld 1967; Bullough 1976). Although the Code of Manu, the first systematic coding of Hindu law, dating from about the first century AD reads: "A man who has committed a bestial crime... shall perform a Samtapana Krikkhra" (Bullough 1976, p. 247), according to Donofrio (1996), in ancient India, the belief in transmigration of souls between animals and humans was combined with acceptance toward bestiality. For example, Kautilya fined a person who copulated with animals only 12 panas, which was much less than for anal intercourse among humans (Bullough 1976). In India too, it was reported that pet dogs and monkeys were kept in harems to service the women (Bullough 1976; Ramsis 1969).

Tantrism often portrays man as a rabbit, bull or horse, and the woman as a doe, mare, or a female elephant, and among the supernatural powers promised to practitioners of various yogic disciplines are those by which a person could become a beast, so that he could have sex with animals and thereby experience sex in its totality (Bullough 1976). In an early legend, Prajapati was said to have cohabited with the dawn goddess Ushas, who tried to escape him by assuming hundreds of different animal shapes. It was through such copulations that all animal species were produced. In Hindu mythology, Mallika, the wife of Prasenajit, used a pet dog for her sexual gratification, and Prasenajit sought satisfaction with a goat (Bullough 1976).

According to the Hindu tradition of erotic painting and sculpture, a human copulating with an animal is actually a human having intercourse with a God incarnated in the form of an animal (Money 1986). Copulation with a sacred cow or monkey is believed to bring good fortune (Edwardes 1959; Christy 1967; Bagley 1968; Rosenberger 1968). During the Hindus' celebrations at the Holi festival, to honor the Goddess Vesanti, open human sexual relations are said to be wildly practiced, and Hindu women are reported to masturbate and perform fellatio on bulls in order to be closer to God (Rosenberger 1968). "Many city youths have their first orgasm dangling from the rump of a sacred cow" (Christy 1967, p. 146), although in an article on sexual problems of adolescence in India, Nagaraja (1983) states that only one percent of the adolescent population suffers from the "abnormal desire" of bestiality (Nagaraja 1983).

Sex shows with dogs, bulls, and water buffalos can be found in the Indian Ocean area and in the Indian Sub-continent (India, Bangladesh, Bhutan, Nepal, Pakistan and Sikkim) (*The Wild Animal Revue* 1994/95, Issue 11). Among the Tamils of Sri Lanka, intercourse with goats and cows is said to be very common (Davis 1954).

Arab Countries, the Middle East, and Africa

According to Rosenfeld (1967) and Rosenberger (1968), the Arabs have long practiced bestiality. They practice bestiality primarily with goats, mares, sheep, sows, asses, and camels, if the latter cooperate. Arab women reportedly have oral sex and intercourse with dogs whenever men are not available to please them (Rosenfeld 1967; Rosenberger 1968). Arab men believe that intercourse with animals increases virility, cures diseases, and enlarges their penises (Dubois-Desaulle 1933; Masters 1962; Bledsoe 1965; Rosenfeld 1967; Bagley 1968; Rosenberger 1968; Gregersen 1983; Dekkers 1994).

The Muslims of Morocco have a similar belief, whereby fathers encourage sons to practice anal and vaginal intercourse with donkeys in order to make the penis grow (Rosenfeld 1967; Rosenberger 1968;

Edwardes and Masters 1977; Dekkers 1994). Boyhood masturbation is scorned in favor of bestiality, and the sight of a group of young Moroccan boys taking turns mounting a donkey is accepted as merely comical (Rosenfeld 1967; Edwardes and Masters 1977). Grown-up men are ridiculed for the practice, but are not punished as long as they perform the act with their own livestock (Masters 1962).

Bestiality is considered better than "zina," which is adultery or fornication. Muslims assume a man has sex with an animal only when he is depraved, or to prevent himself from committing "zina." If discovered, the animal was to be destroyed and eaten (if of an edible species) (Bullough 1976). Dekkers (1994) further reports that Algerian boys have sex with she-asses because marital dowries are so high they cannot afford to get married and are deprived of sex with wives.

Although under Islamic law the penalty for sex with an animal is death (Rosenfeld 1967; L'Etalon Doux 1996), and in ancient times bestiality led to death by stoning of both man and animal (Edwardes 1959; Ramsis 1969), bestiality was tolerated in Islam and widely practiced (Bullough 1976; Dekkers 1994) (interestingly enough, according to Edwardes (1959), Masters (1962), and Ramsis (1969), the Koran makes no mention of sexual relations with animals). A popular Arab saying is that "the pilgrimage to Mecca is not complete without copulating with the camel" (Edwardes 1959; Masters 1962; Bagley 1968).

Among some nomad tribes, intercourse with cattle is still regarded as a ritual of passage for adolescent males (Bullough 1976). Bestiality is found only rarely among the Rwala Bedouins, occasionally in Central Arabia, and frequently among the semi-Bedouins of Northern Israel and Mecca (Masters 1962; Gregersen 1983). It was also reported that as recently as the early part of this century, the nomads' practice of bestiality with their cattle constituted an ordinary feature of pastoral life among the Palestinian Arabs (Masters 1962; Bagley 1968).

During the 1978 American conflict with Iran, the *Little Green Book*, with extracts from the writings of Ayatollah Khomeini, was published. This book contains traditional ritualistically correct views on various issues, among them what to do with a sodomized camel (Gregersen 1983).

Bestiality is common among the Turks (Rosenfeld 1967), who are known for having anal intercourse with mares (Bagley 1968). Some people regard bestiality as sinful only when it involves animals that are edible, such as cattle or sheep. Turks also believe that sex with a donkey makes the human penis grow larger (Gregersen 1983). Today in Turkey, although enforcement of moral laws is very strict, pony, donkey, and dog sex shows are known to run from time to time. The last reported arrests for bestial activity were in 1993 and took place near the Kurdish refugee camps (*The Wild Animal Revue* 1994/95, Issue 11).

There are stories about the notorious side-shows in Aden, Port-Said, Cairo, and Alexandria which offer tourists sex exhibitions of humans and animals (Braun 1967; Rosenberger 1969), and it is reported that brothels in Cairo provide sex shows of women and mule stallions (Hirschfeld 1948; Ullerstam 1966). Egyptian shepherd boys are well known for engaging in sexual relations with animals in their herd. They especially favor fellatio, and rub honey or candy on the penis to encourage the suckling of lambs and goats (Edwardes 1959; Masters 1962; Bagley 1968).

In Lebanon, Beirut was known as a "hot place" for bestiality in the 1960s (*The Wild Animal Revue* 1994/95, Issue 11), and according to Dubois-Desaulle, in 1933 bestiality was still very popular in Syria.

Sexual acts between humans and animals were not punished or even considered socially unacceptable among the Kusai and Masai tribes (inhabitants of Kenya and Tanzania). On the South Sea Island of Kusai, men are reported to use cattle occasionally as sexual objects (Ford and Beach 1951; Dekkers 1994), and Masai male adolescents frequently use female donkeys as a sexual outlet (Ford and Beach 1951; Dekkers 1994; L'Etalon Doux 1996), and as practice, as they believe it improves their lovemaking (Sparks 1977). The Suaheli (Bantu people of Zanzibar/Tanzania) and Arabian fisherman along the coast of Africa, near Mombasa, Kenya, used to believe that unless they had anal sex with the sea-cows they netted, they would be dragged out to sea the next day and drowned by the dead sea-cow's sister. Many people would therefore make the fishermen swear, by the Koran, that they did not have sex with the sea-cow they were selling at the local market (Bledsoe 1965; Love 1992).

At El Yemen, trained baboons were popular sex partners for both men and women, and the women in Abyssinia (Ethiopia) and the Sudan were said to smuggle dog-faced apes (girds) into the harem and have sexual relations with them (Edwardes 1959; Masters 1962; Bagley 1968). Among the Manghabei of Madagascar, bestiality with calves and cows has been observed to be practiced openly by children and adults alike (Masters 1962; Bagley 1968). The people of the Hottentot tribe, nomadic people in southwest Africa, do not consider bestiality to be immoral; they do, however, regard incest in the same negative light as Western people (Rosenberger 1968). Sex with animals used to be a part of the Ibo (Nigerian tribe) male coming-of-age ritual. Every boy had to "successfully" copulate with a specially selected sheep, to the satisfaction of a circle of elders who witnessed the performance. Among the Yoruba (another tribe in Nigeria), there was the custom that a young hunter had to copulate with the first antelope he ever killed, while it was still warm (Gregersen 1983).

Many tribes in Central Africa believe animals to be the ancestors of human beings (Rosenberger 1968). In Voodoo ceremonies, as well as in some other religious and magical rituals, individuals believe themselves as transformed into animals, and have sexual relations either with other humans or with animals of the kind they believe themselves to be (Masters 1962).

South and Central America

The Inca civilization extended down the Pacific coast, from Columbia to Chile and inland to the Andes. In their sexual mores, bestiality was punishable by hanging. Nevertheless, six percent of Inca archaeological decorated specimens, dated from before AD 1000, depict bestiality (Bullough 1976; Tannahill 1992; Dekkers 1994; L'Etalon Doux 1996).

An ancient law in Peru forbade bachelors from having female alpacas in their homes (Mantegazza 1932) because of the many reported cases of bestiality (Maybury 1968; Kullinger 1969). Peruvian men who were unaccompanied by women were further forbidden from herding llamas (Menninger 1951). In South and Central America, bestiality was said to be so prevalent when the Spaniards arrived, that the priests included the "sin of bestiality" in their confessional protocol (Rosenberger 1968; Dekkers 1994).

According to Gregersen (1983), sexual contacts with animals play an important part in the sex life of almost everyone in the Kagaba, an agricultural society in northern Columbia (Gregersen 1983). An ancient pre-Columbian belief among Indians of the Caribbean coast of Columbia relates that adolescent males will not achieve competence in marriage unless they practice intercourse with donkeys (Money 1986).

In a study on the gaucho population living on the border of Brazil and Uruguay, Leal (1989) found the gauchos to understand bestiality as a legitimate practice within a group where the dominant cultural belief consists of mastering the wild. A sexual relationship with certain animals is not only a sanctioned practice within this group, but is seen throughout south Brazil as a herdsmen's or rural tradition. "Barranquear" is the regional term used to refer to male sexual relationship with animals, usually mares. There is a sort of hierarchy of animals to be followed in the "barranqueamento." The sequence starts with the chicken and culminates with the mare. Chickens are for small and young boys, and the act is subject to ridicule. For the gauchos, bestiality shows courage, and the wilder the animal in the animal hierarchy, the more prestigious is the act. Most gauchos do not engage in bestiality as a regular activity, although it is an important part of their sexual initiation.

In the towns and cities of this region, bestiality is considered another form of sexual play among male teenagers. It is tolerated by society as part of growing up and as a necessary erotic experience. Bestiality within this more urban context is practiced with hens, ewes, sows, cows, mules, and mares, but not with cats or dogs. A group of boys will hold the animal while one of them has intercourse with it (Leal 1989). There is no legislation against bestiality in Brazil, either under criminal or civil law. It is an offense only when it is done in a public area (Leal 1989). Brazil is especially known for its sex shows, and some of the latest animal porn films are from this country (*The Wild Animal Revue* 1994/95, Issue 11).

In an analysis of Latin American (Mexican, Cuban) pornography of the 1930s through the 1950s, Di-Lauro and Rabkin (1976) found that bestiality was a common theme. Films such as *Rin Tin Tin Mexicano*, and *A Hunter and His Dog* depict bestiality acts. *The Wild Animal Revue* (1991, Issue 2) further describes a series of 8mm stag films, which appeared during the early 1930s, known as the "Mexican Dog" series. Animal sex shows in Mexico have declined since the days of the 1950s and 1960s, but there are still rumors of the famous donkey shows (*The Wild Animal Revue* 1994/95, Issue 11). Sex shows with animals were common in the brothels of Cuba, but Castro closed down all the brothels (Rosenfeld 1967; Dekkers 1994).

In Balboa, Panama, there used to be night clubs that featured a donkey having intercourse with a woman (Bryant 1982). There has always been an underground trade in bestiality videos and magazines, and United States Customs occasionally checks tapes coming in from Panama (*The Wild Animal Revue* 1994/95, Issue 11).

Native Americans, Canadians and Eskimos

Among Native Americans, bestiality varied from tribe to tribe (Rosenberger 1968). Married men, among the Navaho Indians (in New Mexico, Arizona, and Utah), were known to occasionally engage in bestiality while out herding alone, and unmarried women engaged in bestiality, as well (Deutsch 1948 cited in Donofrio 1996). Bestiality was common among the Crow (native Americans who live in the upper basins of the Yellowstone and Bighorn rivers, in eastern Montana) who had no scruples about having sexual relations with mares and wild animals that had just been killed in the hunt. Although all forms of animal sexual contacts are said to be taboo among the Ojibwa (native Americans and Canadians who live in Michigan, Wisconsin, Minnesota, and Ontario), Ojibwa women were known to have sex with dogs, while Ojibwa men had sexual relations with dogs, bears, moose, beavers, caribou, and porcupines (Gregersen 1983). Cases of bestiality among the Mohave (native Americans who live along the Colorado river in Arizona and California) are known to have involved mares, female asses, heifers, sows, and hens (Menninger 1951). Bestiality is fairly common among the Hopi Indians in north Arizona (Ford and Beach 1951; Dekkers 1994), who consider sex with animals socially acceptable (L'Etalon Doux 1996). Hopi men are reported to have intercourse with burros, dogs, horses, sheep, and chickens (Ford and Beach 1951), and Hopi boys are sometimes directed to animal contacts so they will leave girls alone (Rosenberger 1968; Gregersen 1983). The Sioux (native Americans of the Great Plains) and the Apache (native Americans in south-west US and in north Mexico) had similar views. The Plains Indians (a number of native north American tribes that inhabited the Great Plains, and followed the buffalo) were known to experiment with colts and to use freshly slain animals for sexual purposes (Rosenberger 1968). In the Canadian Indian tribe of the Salteaux, sexual relations between women and dogs are reported. It is also reported that hunters have sex with moose and with female bears they have shot, before the animals get cold (Menninger 1951).

Sexual acts between humans and animals were not punished or even considered socially unacceptable among the Kupfer Eskimos (L'Etalon Doux 1996). Among the Copper Eskimos, intercourse between men and live or dead animals is not infrequent and is not prohibited (Ford and Beach 1951).

The New World — The American Colonies

In Colonial America, a divorce law enacted in 1639, in Plymouth Colony, mentioned bestiality specifically as reason for divorce (L'Etalon Doux 1996), and some sexologists and historians believe bestiality was widespread (Rosenberger 1968; D'Emilio and Freedman 1988). Colonial laws against bestiality required harsh punishment, since the colonists believed these relationships could have reproductive consequences of monstrous offspring. Therefore, the colonists made sure both the person and the animal were executed (D'Emilio and Freedman 1988). The Colony of Pennsylvania ordered life imprisonment and whipping of the person involved in bestiality, at the discretion of the court, and Colonial Virginia law prescribed castration as a remedy for bestiality (Bruno 1984). The first recorded cases of bestiality in the New World took place in 1642 in Plymouth and Massachusetts Colonies. Both men were sentenced to death, and the animals were slaughtered and burned (D'Emilio and Freedman 1988; Bullough 1994; Dekkers 1994).

The United States of America

According to Kinsey, Pomeroy and Martin's1948 study, one American man in about 13 had sexual experience with animals. The authors estimated the number to be eight percent of the male population in the United States. They also stated that animal sexual contacts were largely confined to farm boys, and added that over half of the rural males who had a college education had had some kind of sexual contact with animals (Kinsey, Pomeroy and Martin 1948). Almost four percent of the women interviewed by Kinsey et al. (1953) reported having had sexual contact with animals after they had become adolescents.

Rumors about orgies involving animals among swinging circles have been reported (Bagley 1968; Bryant 1982), and according to Dumont (1970) there used to be a guest ranch in Texas, as late as 1970, which arranged sexual relations between the guests and various horses trained for performing sexual acts. The Pet Book series, detailing bestiality fantasies, from Greenleaf Classics in San Diego, California, has flourished since the early 1970s (*The Wild Animal Revue* 1991, Issue 1). A full length, underground movie was reportedly shown in some San Francisco adult movie theaters about 20 years ago. The film was called *Animal Lovers* and portrayed the female star engaging in intercourse with various types of animals including a dog, a donkey, and a pig (Bryant 1982). There are also the Color Climax' 8mm animal films, such as *Dog Fuckers*, *Horse Lovers*, and *Horsepower*, all from 1970. Another two 8mm stag films appeared in the early 1970s in which porn star, Linda Lovelace, had sex with a large dog. Lovelace, however, has denied her participation in such films (*The Wild Animal Revue* 1991, Issue 1). There have been reports of underground, private, local, animal shows in the United States and Canada, but nothing organized. At one mid-Western high school, the football team still gets a "goat show" after "home coming," reportedly, a tradition for over 20 years. There were also reports about some wild animal sex shows during construction of the Alaskan pipe line (*The Wild Animal Revue* 1994/95, Issue 11).

In 1962, Illinois became the first American state to revise its criminal code, and oral-genital contacts, anal intercourse between consenting adults in private, and sexual acts with animals, were no longer considered criminal offenses (L'Etalon Doux 1996). In 1997, twenty-five states, the District of Columbia, and the United States Government outlawed bestiality. The sentences ranged from a mere fine of not more than $500 in Tennessee to an indeterminate life sentence in Michigan (Miletski 2002). The laws in the United States have been changing, and according to Richard (2001), in 2001, three states—Iowa, Maine, and Oregon—passed laws criminalizing bestiality.

The late Mark Matthews, in his book *The Horseman* (1994), helped increase awareness of the existence of "zoophilia" (i.e., in addition to engaging in bestiality, the zoophile, or "zoo," also feels love and sexual attraction toward the animals). In recent years, the Broadway theater has been increasingly open in its portrayal of the full spectrum of sexual themes and activities. Productions such as, *Futz* (Bryant 1982) and the very recent *The Goat*, have depicted the themes of bestiality and zoophilia.

The Internet

Alt.sex.bestiality (A.S.B.) was one of the first Internet news groups which started around 1990 as someone's idea of a joke. Soon, A.S.B. grew and matured into a discussion and support group (Fox 1994; Andriette 1996; Montclair 1997), providing information about health issues, laws governing bestiality, bibliographic references, "how to" guides, written, and pictorial erotica, and information about the "zoo community's" events (Donofrio 1996). Most people in this newsgroup had sexual relations with animals, and many were quite proud of it. There were also many others who have not had any sexual contact with animals, but who were eager to do so (Fox 1994). According to Andriette (1996), for most "zoos," finding other "zoos" has changed their lives. It has given them a new self understanding, and connected them with like-minded friends (Andriette 1996).

Stasya's Home Page, which was about zoophilia, was initiated in September of 1995, and averaged a "hit" (a visitor) every three minutes. Stasya reported (in 1996) receiving anywhere from two to six messages per day from people saying "thank you for being there" (Stasya 1996). In January of 1999, according to the Humane Society of the United States' web site, in one Internet search using the term "bestiality," it found 85,771 documents. These days, one can find numerous web sites, chat rooms, and pet forums exclusively devoted to bestiality, zoophilia, and related pornography online. And Byrd (2000) points out that bestiality/zoophilia has never been more present, even in fashion magazine advertisements and television commercials.

Conclusion

A more in-depth exploration of the history of bestiality is beyond the scope of this paper. However, it is apparent and important to acknowledge that man has engaged in bestiality since the dawn of civilization, in almost every culture and place in the world. Although individuals have been punished, sometimes tortured and killed, for engaging in bestiality, the behavior and the pre-occupation with bestiality has persisted. Even more important are the reports that bestiality is still an integral part of many people's lives, whether in myth, art, literature, or as actual sexual behavior. Although no one knows how prevalent sexual behavior with animals is, bestiality is unquestionably among us.

Acknowledgements

I would like to take this opportunity to thank Andrea Beetz, Ph.D., for inviting me to write this article, and to share my knowledge/study with my colleagues through this distinguished journal. I would also like to thank my husband, Arthur Goldberg, for his editorial assistance, and the two reviewers and their thoughtful suggestions.

References

Allen, C. 1979. *The Sexual Perversions and Abnormalities—A Study in the Psychology of Paraphilia.* 2nd edn. Westport, CT: Greenwood Press.

Andriette, B. 1996, March. Laying with beasts. *The Guide.*

Bagley, H. L. 1968. *The Beast Seekers.* Atlanta, GA: Pendulum Books.

Blake, R. 1972. *Beauty/Beast: A Study of Animal Eroticism*; Vol. 2. 2nd edn. San Diego, CA: Greenleaf Classics, Inc.

Bledsoe, R. J. 1965. *Male Sexual Deviations and Bizarre Practices.* 2nd edn. Los Angeles, CA: Sherbourne Press, Inc.

Bloch, I. 1933. *Anthropological Studies in the Strange Sexual Practices of All Races in All Ages.* New York: Anthropological Press.

Braun, W. 1967. *The Cruel and the Meek.* New York, NY: Lyle Stuart, Inc.

Bruno, R. A. 1984. Laws governing sexual behavior in the United States. Master's Thesis. San Francisco, CA: The Institute for Advanced Study of Human Sexuality.

Bryant, C. D. 1982. *Sexual Deviancy and Social Proscription.* New York: Human Sciences Press, Inc.

Bullough, V. L. 1976. *Sexual Variance in Society and History.* Chicago, IL: The University of Chicago Press.

Bullough, V. L. 1994. Foreword. In *The Horseman; Obsessions of a Zoophile*, 5–9, M. Matthews. Amherst, NY: Prometheus Books.

Byrd, M. 2000. The joy of beasts. *The Independent (The Sunday Review)*, 3 December.

Christy, R. 1967. *The Animal Lovers.* Cleveland, OH: Classics Library.

Cornog, M. and Perper, T. 1994. Bestiality. In *Human Sexuality: An Encyclopedia*, 60–63. ed. V. L. Bullough and B. Bullough. New York: Garland Publishing, Inc.

Davies, C. 1982. Sexual taboos and social boundaries. *American Journal of Sociology* 87(5): 1032–1063.

Davis, P. 1954. *Sex Perversion and the Law, Volume One.* 5th edn. New York: Mental Health Press.

Dekkers, M. 1994. *Dearest Pet: On Bestiality.* 2nd edn. Translated by P. Vincent. New York: Verso.

D'Emilio, J. and Freedman, E. B. 1988. *Intimate Matters: A History of Sexuality in America.* New York: Harper & Row, Publishers.

Deutsch, A. ed. 1948. *Sex Habits of American Men: a Symposium on the Kinsey Report.* New York: Prentice Hall.

Di-Lauro, A. and Rabkin, G. 1976. *Dirty Movies - An Illustrated History of the Stag Film* 1915–1970. New York, NY: Chelsea House.

Donofrio, R. 1996. Human/animal sexual contact: A descriptive-exploratory study. Doctoral dissertation. San Francisco, CA: The Institute for Advanced Study of Human Sexuality.

Douglas, A. 1992. *The Beast Within.* New York, NY: Avon Books.

Dubois-Desaulle, G. 1933. *Bestiality: An Historical, Medical, Legal, and Literary Study.* Translated by A. F. N. New York: The Panurge Press.

Dumont, A. 1970. *Wealthy Women and Bestiality*. North Hollywood, CA: Barclay House.

Edwardes, A. 1959. *The Jewel in the Lotus.* Julian Press, Inc.

Edwardes, A. and Masters, R. E. L. 1977. *The Cradle of Erotica*. New York, NY: Bantam Books.

Ellison, A. 1970. *Sex Between Humans & Animals: The Psycho-Mythic Meaning of Bestiality*. San Diego, CA: Academy Press.

Evans, E. P. 1987. *The Criminal Prosecution and Capital Punishment of Animals.* 2nd edn. London: Faber and Faber.

Ford, C. S. and Beach, F. A. 1951. *Patterns of Sexual Behavior.* Harper & Brothers, Publishers.

Fox, R. 1994. *The alt.sex.bestiality Frequently Asked Questions* [online]. Redvane@atheling.demon.co.uk.

Friday, N. 1981. *Men in Love.* 2nd edn. New York, NY: Dell Publishing.

Greenland, C. 1983. Sex law reform in an international perspective: England and Wales and Canada. *Bulletin of the American Academy of Psychiatry and the Law* 11(4): 309–330.

Greenwood, M. D. 1963. *Unusual Sex Practices.* Hollywood, CA: The Genell Corp.

Gregersen, E. 1983. *Sexual Practices: The Story of Human Sexuality.* New York: Franklin Watts.

Haeberle, E. J. 1978. *The Sex Atlas.* New York, NY: The Seabury Press.

Hamilton, D. 1981. *Bestial Desires.* South Laguna, CA: Publisher's Consultants.

Harris, E. 1969. *Animals as Sex Partners*. Los Angeles, CA: Ultima House.

Hirschfeld, M. 1948. *Sexual Anomalies.* New York: Emerson Books, Inc.

Home Office. 2002. *Protecting the Public—Strengthening Protection Against Sex Offenders and Reforming the Law on Sexual Offences.* Norwich, U.K: The Stationery Office.

Hunt, M. 1974. *Sexual Behavior in the 1970s*. Chicago, IL: Playboy Press.

Kinsey, A. C., Pomeroy, W. B. and Martin, C. E. 1948. *Sexual Behavior in the Human Male.* Philadelphia, PA: W. B. Saunders Company.

Kinsey, A. C., Pomeroy, W. B., Martin, C. E. and Gebhard, P. H. 1953. *Sexual Behavior in the Human Female.* Philadelphia, PA: W. B. Saunders Co.

Kullinger, J. L. 1969. *A Very Good Book About Some Very Bad People — The Arrogant Sexuals.* Las Vegas: Nevada: M-T Publishers, Inc.

Leal, O. F. 1989. The Gauchos: Male culture and identity in the Pampas. Doctoral dissertation. Berkeley: University of California.

L'Etalon Doux. 1996. *Zoophilia and the Law: History* [online]. An63019@anon.penet.fi.

Liliequist, J. 1988. Peasants against nature—Crossing the boundaries between man and animal in 17th and 18th century Sweden. Paper presented at the conference "Man and the Animal World," Nijmegen, Holland, September, 1988.

Love, B. 1992. *Encyclopedia of Unusual Sex Practices.* Fort Lee, NJ: Barricade Books, Inc.

Mantegazza, P. 1886. *L'Amour dans l'Humanite.* Paris.

Mantegazza, P. 1932. *Anthropological Studies of Sexual Relations of Mankind.* New York: Anthropological Press.

Masters, R. E. L. 1962. *Forbidden Sexual Behavior & Morality.* New York, NY: Lancer Books, Inc.

Masters, R. E. L. 1966. *Sex-Driven People.* Los Angeles, CA: Sherbourne Press, Inc.

Matthews, M. 1994. *The Horseman; Obsessions of a Zoophile.* Amherst, NY: Prometheus Books.

Maybury, D. O. 1968. *The Secret Perverts.* Los Angeles, CA: Classic Publications.

Menninger, K. A. 1951. Totemic aspects of contemporary attitudes toward animals. In *Psychoanalysis and Culture: Essays in Honor of Geza Roheim*, 42–74, ed. G. B. Wilbur, and W. Muensterberger. New York, NY: International Universities Press.

Miletski, H. 2002. *Understanding Bestiality and Zoophilia.* Bethesda, MD: East West Publishing, LLC.

Money, J. 1986. *Lovemaps.* New York: Irvington Publishers, Inc.

Montclair, R. 1997. Tails of bestiality. *Black Book Magazine.*

Monter, E. W. 1981. Sodomy and heresy in early modern Switzerland. *Journal of Homosexuality* 6(1-2): 41–55.

Nagaraja, J. 1983. Sexual problems in adolescence. *Child Psychiatry Quarterly* 16(1): 9–18.

Niemoeller, A. F. 1946a. *Bestiality in Ancient and Modern Times.* Girard, Kansas: Haldeman-Julius Publications.

Niemoeller, A. F. 1946b. *Bestiality and the Law: A Resume of the Law and Punishments for Bestiality with Typical Cases from Fifteenth Century to the Present.* Girard, Kansas: Haldeman-Julius Publications.

Ramsis, J. A. 1969. *Casebook: Animal Love.* Pendulum Books, Inc.

Richard, D. 2001. Forbidden love: viewing the question of animal love from two sides. *Contemporary Sexuality* 35(10): 1, 4–7.

Rosenberger, J. R. 1968. *Bestiality.* Los Angeles, CA: Medco Books.

Rosenfeld, J. R. 1967. *The Animal Lovers.* Atlanta, GA: Pendulum Books.

Salisbury, J. E. 1994. *The Beast Within—Animals in the Middle Ages.* New York: Routledge.

Somers, J. E. 1966. *Sick Sex.* Hollywood, CA: Venice Publishing Corporation.

Sparks, J. 1977. *The Sexual Connection — Mating the Wild Way.* New York, NY: McGraw-Hill Book Company.

Stasya. 1996. *Stasya's Home Page* [online]. Accessed at http://www.av.qnet.com~stasya.

Stekel, W. 1952. *Patterns of Psychosexual Infantilism.* New York: Liveright Publishing Corporation.

Tannahill, R. 1992. *Sex in History.* Rev. edn. Scarborough House/Publishers.

Taylor, T. 1996. *The Prehistory of Sex.* New York: Bantam Books.

Trimble, J. F. 1969. *Female Bestiality.* Torrance, CA: Monogram Publications, Inc.

Ullerstam, L. 1966. *The Erotic Minorities.* New York, NY: Grove Press, Inc.

Waine, W. W. 1968. *Canine Sexualis.* San Diego, CA: Publisher's Export Co., Inc.

Sexual relations with animals (zoophilia): An unrecognized problem in animal welfare legislation

Gieri Bolliger and Antoine F. Goetschel

Foundation for the Animal in the Law, Bern and Zurich, Switzerland

Abstract

Sexual contact with animals (zoophilia) has always been part of human culture, and is more widespread than generally thought. While it was severely penalized on ethical and religious grounds for centuries, the age of Enlightenment led to more rational views on this topic and, consequently, milder punishment, until eventually the sanctions were lifted in most countries. Apart from presenting a summary about the changes in legislation dealing with zoophilia over time, this article provides an outline of today's legal position on zoophilia throughout the world. A closer look at the existing laws reveals loopholes in the criminal codes and animal welfare legislations regarding the punishment of zoophilic acts. Taking into account the "dignity of the animal" should be an important factor in the revision of existing laws. This will be discussed.

Keywords: *animal abuse, animal welfare, bestiality, law, legislation, zoophilia*

Sodomy, Zoophilia, Bestiality and Other Terms

Sexual contact between human beings and animals has a long history. Over the course of this history, not only the general perception but also the naming of such deeds has changed. In the German colloquial language, sexual contact between human beings and animals is generally called "sodomy." This term stems from the biblical city of Sodom on the Dead Sea, the people of which were infamous for their excessive and depraved lifestyle—namely for their distinct tendency towards various forms of fornication. Until recently, the term sodomy was not used exclusively to distinguish sexual intercourse with animals; it was also used for any "unnatural fornication," which means any sexual practice not serving the purpose of procreation. In many cultures today, this is still so. According to the Catholic Church, some other sexual acts, such as coitus

Address for correspondence: Dr. Antoine F. Goetschel, Foundation for the Animal in the Law, Wildbachstrasse 46, Postfach 1033, CH-8034 Zürich, Switzerland. Ph: +41 (0) 43 443 06 43; fax: +41 (0) 43 443 06 46; e-mail: info@tierimrecht.org; www.animallaw.ch

with the Devil or any non-Christian and even certain acts in marriage (e.g., coitus in "unnatural positions," masturbation or "pollution," which means the very first ejaculation), were called sodomy. Moreover, the term "sodomy" can be misunderstood, and is not suited for the sole description of intimacy with animals because in many languages (e.g., English and French) it stands primarily or even exclusively for same-sex practices between human beings.

Occasionally, sexual contact with animals is paraphrased in terms such as "fornication with animals" or "bestiality." Today, the scientifically correct term of "zoophilia" should be used (Rosenbauer 1997; Beetz 2002; Miletski 2002). And although the literal translation simply means "affection" or "love" of animals (Hunold 1970; Rosenbauer 1997; Hoffmann 2003), it expressly denotes not a "normal" love for animals in terms of a mere emotional, platonic, devotion. Rather, it refers to a strong, erotic relationship with an animal, in such a manner that it leads to its inclusion in sexually motivated and targeted acts, with the direct intention of sexually arousing one-self, the animal or another party. Against this background, one can distinguish between five sexual acts between human beings and animals (Massen 1994): genital acts (anal and vaginal intercourse, insertion of fingers, hands, arms or foreign objects), oral-genital acts (fellatio, cunnilingus), masturbation, frotteurism (rubbing of the genitalia or the entire body on the animal) and voyeurism (observation by third parties during sexual interactions with animals). Acts which are not sexually motivated, such as the petting or hugging of animals, riding, and any conscious or unconscious fantasies of zoophilic acts (Beetz 2002; Friday 2004) or the mere observation of intercourse between animals do not fall under the umbrella of "zoophilia" (Kinsey, Baacke and Hasenclever 1963; Muth 1969).

Zoophilia is a phenomenon which is ultimately as complex as sexuality itself and comprises just as many varieties. Therefore the boundaries between sub-groups (such as zooerasty, zoostuprum, zoofetishism etc.) can be blurred. For the sake of clarity, we shall abstain from further differentiations, with one exception, that being to distinguish between violent and non-violent zoophilia. As not all animals comply with the humans' wishes and do not let sexual intercourse occur, it is frequently effected by using physical force. If the sexual tension does not find immediate release, it occasionally leads to acts of destruction, for which mechanical instruments such as pitchforks, broomsticks or tapered sticks are sometimes used (Merki 1948; von Hentig 1962; Berg 1963; Muth 1969; Stettner 1990). The results of this abuse are often grave injuries, even the death of the animal (Weidner 1972). If it is sexually stimulating or even satisfactory for a person to inflict pain to an animal, to mutilate or even kill it, it is called "zoosadism" (Weidner 1972), of which again several types are known (von Hentig 1962; Masters 1965). For example, there is the type called "cattle stabber." This is a person who kills cows, horses, sheep and goats in the context of his sexual deeds (Rosenbauer 1997). Other offenders strangle chickens, geese and ducks during the sexual encounter, or cut their throats, in order to stimulate themselves by watching the dying animal twitch (Dekkers 1996; Hoffmann 2003). In many places, specialized prostitutes allow their clients to involve animals in intimate acts to torture and kill them (Hunold 1970, Rosenbauer 1997)—the animals are brought along by the client or they are made available. A further form of zoosadism involves setting sexually excited animals on girls or women (Masters 1965; Hunold 1970). However, it remains to be said that not every slaying of an animal after a zoophilic act necessarily springs from a sadistic drive. Some animals die as a result of unwanted "accidents" or are killed in the aftermath of the deed by the perpetrator, arguably because of their disgust about their own actions (Muth 1969).

Zoophilia Over Time

Sexual interactions between human beings and animals are an integral part of almost all cultures and religions, and have been considered a primal phenomenon of human society (Muth 1969; Miletski 2002). According to numerous traditions, zoophilia appears to have played a significant role amongst ancient peoples, as well (Masters 1965). Thus, many totem cults trace back to the concept of an animal ancestor (Rosenbauer 1997) and the sexual union between a human and an animal, resulting in the creation of mixed creatures (von Hentig 1962; Massen 1994; Dekkers 1996; Guggenbühl 2002).

Prevalence of Zoophilia

While most sexual practices have been described and discussed in print media and on film and television, zoophilia largely remains a social taboo (Massen 1994). Media reports on the topic are scarce (but see Illi 1998, for an example), and even animal welfare organizations seem to address zoophilia only reluctantly.

Reliable research or statistics on the prevalence of zoophilic contacts in the total population are lacking. And despite that sex with animals isn't illegal in many countries, an affected person is not likely to admit publicly and voluntarily to their deed. Hence, the actual prevalence of zoophilia can only be estimated. Confounding the situation is that many, if not most, cases remain undiscovered.

Because of this lack of current, accurate statistics, we need to fall back on the data of the Kinsey Reports of the mid-20th century, which impressively disproved the assumption that sexual acts with animals were a rare phenomenon in society. The studies, conducted by zoologist and social researcher Alfred C. Kinsey and his associates between 1938 and 1947, were based on 20,000 interviews dealing with the sexual behavior of North American men and women. This research brought to light that 8% of the male and 3.5% of the female populations of the US had had at least one zoophilic encounter during their lives. Among the rural population, which had easy access to animals, 17% of the men surveyed gave accounts of intimate experiences with animals, leading to orgasm (Kinsey et al. 1955). In some communities, estimates of up to 65% were determined (Kinsey et al. 1955). In the urban population, however, the percentage was a much lower: up to 4%, and zoophilic contacts took place mainly during temporary sojourns to the country (Kinsey et al. 1955). Although the research indicated that about eight million US citizens had engaged in zoophilia, this can only be a mere fraction of the actual number, as one has to take into account that many people would conceal such experiences because of it being an illegal, punishable offense at the time of the surveys. As well, moral reasons and a sense of shame would have influenced the results.

Although the Kinsey research is sixty years old and its generally blurred terminology has often been erroneously interpreted and misunderstood (Rosenbauer 1997), its data are still considered the most comprehensive available today. Subject to the aforementioned reservations and the certainty that many factors have changed in the meantime, the study allows us to draw at least certain conclusions about today's situation. Thus, one can assume that zoophilic actions today—in America and in Europe—are just as widespread as they were in Kinsey's time. However, it has to be recognized that today there are far more pets than in earlier times (Goetschel and Bolliger 2003), and thus access to animals is easier for urban people (Kinsey et al. 1955; Masters 1965; Massen 1994). Therefore not only the total number of sexual relations with animals can be assumed to be much higher, but one would expect more women to be involved, as well (Dekkers 1996).

Because the topic of zoophilia is persistently hushed-up and there is a paucity of literature on it, it is understandable that it is believed to be a rare phenomenon. But the many advertisements in notorious magazines, as well as the overwhelming amount of zoophilic material that can be found on the Internet, prove that this is a false conclusion, and indicates the existence of an underground "scene" (Beetz 2002).

Legal Discussion

Historical Abridgement

From a judicial point of view, zoophilia has been judged in various ways over time (Merki 1948; Masters 1965; Muth 1969; Dekkers1996). Possibly the first ever mention of it can be found in the Hittite compendium of laws (Muth 1969; Massen 1994), which dates back to about 1300 BC, where sexual acts committed by men—but not by women (Kinsey, Baacke and Hasenclever 1963)—with certain animals (e.g., cattle, sheep, pigs or dogs) were prohibited under penalty of death (Kinsey et al. 1955; von Hentig 1962; Masters 1965; Massen 1994; Dekkers 1996). This is remarkable, especially considering that other crimes such as manslaughter, arson or battery were only penalized with fines (Massen 1994).

After mono-ethical Judaism crowded out religious animal cults in all societies, Mosaic legislation contained explicit bans on zoophilia (Hunold 1970). In the Old Testament, as well as in the Talmud, zoophilia is considered disrespectful of divine Creation (Weidner 1972) and any party to such an act—man, woman, animal—invariably was punished with death by stoning (Merki 1948; Krings 1995; Dekkers 1996). The Talmud even outlawed any woman from being alone with an animal, in order to rule out any suspicion from the outset (Muth 1969; Christy 1970).

Today, most countries no longer consider zoophilia sacrilegious. However, it remains outlawed in some places, especially in countries with Germanic and Anglo-American legislation. For example, the single states of Germany, Austria and a large proportion of the Swiss cantons have adhered to treating zoophilia as a crime, as has England and the North American and Scandinavian states (Merki 1948). The reasons behind the retention of the bans, which was dependent on the opinion of the each legislator, are that sexual contact with animals generally arouses disgust and leads to "disdain of marriage," "endangering of family life" or "de-population" and to a "deterioration" of the state (Muth 1969; Weidner 1972; Dekkers 1996). Also, human dignity was an important issue; it was gravely wounded by any sexual contact with an animal. Indeed, this kind of sexual misconduct was considered such a fundamental breach of duties that the perpetrator not only relinquished his/her own dignity, but also harmed the socially important awareness of self-worth of all mankind (Grassberger 1968).

Today's Legal Position

Modern criminal law makes a clear distinction between law and ethics, and leaves the punishment for moral transgression exclusively to society (Ford and Beach 1954; Dekkers 1996; Arnold, Eysenck and Meili 1997; Hoffmann 2003). With this in mind, it is perhaps not surprising that in 1950 zoophilia was exempt from punishment in over 80% of European states (Muth 1969).

Over the last century, Scandinavian countries followed their Roman examples and deleted the offenses of sexual contacts with animals from their criminal code (Stettner 1990). As a rule, acts of zoophilia are only relevant to law if they cause cruelty to animals, and this has become part of almost all national animal welfare legislation: if animals demonstrably suffer pain or injury in the context of sexual acts, the deed has to be prosecuted. In some countries, bans on sexual acts in front of minors can be found (Stettner 1990), while only very few European countries strictly prohibit the production, distribution and the possession of zoophilic pornographic material. In Italy, Spain and the Netherlands, for example, such material is legally and publicly available in erotica shops.

Laws Across the World

German and English-Speaking States

The only countries to have adhered to their fundamental bans on zoophilia are those of Germanic and the heavily influenced Anglo-American legislations. However, a variety of legal definitions exist within those legislations, as is demonstrated by the overview which follows. The legal status of the specified countries is comparable to the Swiss legal status in many areas, which is why repetitions have been largely omitted and the elaboration focuses on specific differences.

Switzerland

Since the coming into effect of the federal criminal code (StGB/CH) on 1 January 1943, Switzerland has maintained a standardized national criminal law. Up until this point, it had been left up to the individual cantonal legislators to rule on acts of zoophilia (Merki 1948). The cantonal criminal code was divided into three groups which differed fundamentally from each other. While French- and Italian-speaking cantons relied heavily on the idea of Enlightenment and therefore did not prosecute any sexual acts with animals *per se*, such acts were, by moral considerations, still outlawed under threat of punishment in most German-speaking cantons (Merki 1948; Krings 1995). Not all acts that offended morals and caused shame were outlawed, but rather only those where there was physical contact with the genitalia of an animal, with the intention to trigger sexual stimulation or satisfaction in oneself or the animal (Merki 1948). The severity of the punishment depended foremost on the "similarity to coitus" of the crime. Over time the interpretation of offense for "unnatural fornication" underwent several changes (Merki 1948; Muth 1969). Finally, it was determined that it was an offense for any person to place their genitals on the body of an animal and simulate coitus; it was not required for the person to engage in actual intercourse or to reach sexual climax. Any act that was solely intended to stimulate or satisfy the animal remained exempt from punishment (Muth 1969). The third group of cantons consisted of a combination of Romanic and Germanic law. Here, sex with animals was only punishable if it invoked a public offense (Merki 1948).

The creation of the Swiss Penal Code (Strafgesetzbuch, StGB/CH) aimed to unify Swiss criminal law and adjust the conflicting cantonal positions. In fact, in 1942 seven cantons made use of the authorization Art. 335 StGB/CH and incorporated the offense of "fornication with animals" into their penal law. Basle-Country, Thurgan and Appenzell Inner-Rhodes were the only cantons to adhere to this, such that today zoophilia still incurs a penalty in these cantons. However, no sentences related to zoophilia have been passed there for decades.

The federal animal welfare legislation that came into force in 1981 (TSchG/CH) does not contain an explicit ban on zoophilia (Vogel 1980). According to Art. 2, an animal must not be exposed to unjustified pain, suffering, injury or fear (Goetschel 1986). Actual penal regulations can be found in Art. 27–29, where, again, sexual actions with animals are not explicitly forbidden (Merki 1948; Vogel 1980). At best, the offense of cruelty to animals applies. If an animal is demonstrably abused, stressed or willfully and inhumanely killed in the context of sexual acts, the perpetrator can be penalized with a prison sentence of up to three years or a fine of up to 40,000 Swiss francs. In a case of negligence, the penalty is either confinement or payment of a fine (Goetschel 1986).

Because zoophilia is basically exempt from punishment, the conscious supplying, lending or selling of animals for such purposes is not prohibited. Therefore, it is also legal to train and accustom animals to sexually targeted deeds. Such acts, again, can only be considered punishable if they meet the requirements of an offense under Art. 27.

Despite zoophilia not being considered an offense in Swiss legislation (with the exception of the cantons of Basle-Country, Thurgan and Appenzell Inner-Rhodes), a zoophilic act conflicts with the federal penal code. If an animal belonging to another person is used, for example, the offense of damage to property according to Art. 144 StGB/CH may apply. According to this clause, a person damaging, destroying or rendering inoperative other people's property can be penalized with up to three years in prison or a fine. According to Swiss legislation, as of the beginning of April 2003, animals are no longer considered mere things (Goetschel and Bolliger 2003).

Because a new article was inserted in the Swiss penal code (Art. 110 Ziff. 4bis) at the same time, Art. 144 StGB/CH also applies when an animal is injured or killed, if the perpetrator is not the owner. Contrary to the earlier-mentioned offense of cruelty to animals, according to Art. 144 StGB/CH, damage to property, injury to or killing of an animal are only punishable by law if committed deliberately and only then if the owner of the animal makes a claim. The negligent injury or killing of another's animal thus remains just as exempt from punishment as the deliberate act when the owner consents to such a deed, or subsequently decides against initiating legal proceedings (Goetschel and Bolliger 2003). As is the case with the offense of cruelty to animals, the requirements for an offense, according to Art. 144 StGB/CH, are already fulfilled, however, if the animal is hurt (even slightly). This means that in order to make a claim for damage of property, neither extensive pain, suffering nor injury have to be inflicted on the animal.

Germany

The German "Reichs-Criminal Code" (RStGB) of 1871 outlawed zoophilia in § 175. After its creation, the relevant paragraph was greatly disputed, mainly because it outlawed male homosexuality, as well. In 1935, these two crimes were retained under the national socialist legislation; however, they were divided by content. From then on "fornication with animals" was recorded in § 175b RStGB, and penalized with up to five years in prison.

After long-term political discussions about whether or not zoophilia was to be retained in the criminal code (Weidner 1972), in 1969 the relevant paragraph was deleted without substitution. Up until then, annually around 200 people were convicted of sexual acts with animals (Weidner 1972; Rosenbauer 1997; Beetz 2002; Hoffmann 2003). After 1st April 1970, any such deed was no longer prosecuted by law in Germany. Reasons for the abolition of the regulation included its marginal meaning in the judicial practice, and that most perpetrators had abnormal affinities by nature and thus were not fearful of punishment. It was also argued that there were no criminal political motives for the retention of this penal norm. That a person debases him/herself by committing a zoophilic act did not provide sufficient cause for legal sanctioning, and observations that zoophiles would go on to commit other sexual offenses later on, did not justify the culpability of fornication with animals either. Furthermore, it was reasoned that punishment for the offense in the context of cruelty to animals would be considered if an animal had suffered distress or rough handling during any sexual act, and that the animal, as a piece of property, was sufficiently protected by the existing criminal regulations dealing with damage to property (Weidner 1972; Frey 2003).

As is the case in Switzerland, zoophilic acts are threatened by penalty in Germany if they involve obvious cruelty to animals. According to § 17 TierSchG, any person who either inflicts substantial pain or suffering, or causes continuing or repetitive pain or suffering, could be fined or imprisoned for up to three years. The same threat of punishment applies to the killing of an animal in the context of a zoophilic action, because such an action is not regarded as a reasonable cause for animal slaughter, according to the animal welfare legislation (Ort and Reckewell 2002). If an act of zoophilia cannot be penalized for lack of any of the aforementioned reasons, an administrative offense, according to § 18 TierSchG, may pertain. According to this article, a person commits an offense if they inflict significant pain or suffering without good reason to any animal in their possession or in their voluntary or imposed care. Both the deliberate and negligent acts are punishable, if the animal's owner commits them. However, if a third party causes an animal significant harm, an intentional or eventually intentional act is required in order for such a deed to be punishable. Therefore, it has to be proven that the person inflicted the injuries knowingly and willingly on the animal, or at least thought about them and accepted the outcomes. Such an administrative offense is penalized with a maximum fine of 25,000 Euro; it is not prosecuted by a court, but rather by an administrative authority (Stettner 1990; Ort and Reckewell 2002; Frey 2003; Buschmann 2004).

In Germany, as is the case in Switzerland, many deeds related to zoophilic actions find their legal boundaries in the national criminal code. According to § 303 StGB/D, a person who deliberately and unlawfully injures or kills another's animal can be fined or imprisoned for up to two years, if the aggrieved owner makes a claim for damage of property. If need be, offenses of trespassing (according to § 123 StGB/D—"fence-hopping") or indecency (if a zoophilic act is committed in public—§ 183a StGB/D) may also apply. Additionally, any person who produces or distributes zoo-pornographic material can be fined or imprisoned for up to three years. Contrary to Swiss legislation, German law allows for the purchasing and owning of such material.

Austria

According to § 130 of the code of law issued in 1852 in Austria, and remaining in effect into the 1970s, sexual acts with animals were illegal and the perpetrator could be imprisoned for up to five years. Not only acts similar to coitus were punished under this law: it was also an offense if a person's genitals were placed anywhere on a living animal with the intention to gain sexual satisfaction (Merki 1948).

Based on this regulation, about 50 people were convicted annually. However, in 1971 it was abolished without substitution because of criminal, political and theoretical legal considerations. It was argued that zoophilic "aberrations" did not pose a threat to the general public; in most cases only involved adolescents; and that social rather than legal mechanisms would be more effective in preventing its spread. Furthermore, it was argued that a threat of punishment and the enforcement of a sentence would have hardly any discouraging effect.

A norm for how zoophilic actions should be penalized cannot be found in the Austrian criminal code (StGB/Ö), in the animal welfare legislations of the nine individual provinces, or in the new uniform national animal welfare legislation, which came into effect on 1st January 2005 (TSchG/Ö). The Austrian criminal code only prohibits the advertising of fornication with animals in its § 220a StGB/Ö: whoever uses print or film or any other medium to publicly appeal for sex with animals, or approves of such an act in a suggestive manner to a third party, can be imprisoned for up to six months or be given a fine of up to 360 day's rates[1].

The offense of cruelty to animals was also governed by the criminal code until the beginning of 2005 (Goetschel and Bolliger 2003). According to § 222 StGB/Ö, an offense occurs if an animal is "abused brutally" or "inflicted with unnecessary agony." Such an offense can incur a prison sentence of up to a year or a fine of 360 day's rates. Again, a zoophilic act is only sanctioned if the animal suffers from substantial pain, injury or distress. The offense of cruelty to animals is carried over into the new animal welfare legislation of 1st January 2005 (TSchG/Ö). Any violation of this regulation will be punished with a fine of up to 7,500 Euro, or 15,000 Euro in cases of recurrence.

The production and trade of any zoophilic images is regulated under the national law on pornography. Under this law, the manufacturing, publishing, importing and exporting, and distributing of obscene writings and images are considered crimes, provided that they are carried out with commercial intentions. The penalty for these offenses is a prison sentence of up to a year and an additional fine of 360 day's rates (§ 1 Pornografiegesetz). As in Germany, the mere possession of zoo-pornographic material is legal in Austria. The same applies for the passing on or making available of such material, as long as it is not done with commercial intentions and is not consciously handed on to teenagers under the age of 16. Otherwise, according to § 2 of the law on pornography, a prison sentence of up to six months or a fine of up to 360 day's rates can be imposed.

Great Britain

Up to the middle of the last century, British law stood out for its severe penalties for people who engaged in sexual acts with animals. The so-called "consummated fornication" with animal, for which the slightest penetration by the male genitals sufficed, was an offense which incurred a prison sentence from ten years to life. Also, the mere attempt at such an act, as well as other zoophilic actions, incurred prison sentences (Merki 1948).

In § 12 Sexual Offences Act, established in 1956, zoophilia is clearly defined as a punishable deed. The so-called "buggery" included any intercourse "in any manner between man or woman and beast," and up until recently, perpetrators were threatened with a life sentence; a conviction required the participation of jurors in a trial (Stettner 1990). The crown law committee addressed the aforementioned regulation, among other issues, in a report about sexual offenses published in 1984. They concluded that zoophilia was to remain a punishable act and that an abolition of the existing regulation would be construed as general tolerance of such practices, which would bring about the danger of an increase in such offenses. However, the committee recommended judging this offense in summary proceedings, without the participation of jurors. The committee also suggested reducing the maximum penalty to six months in prison; however, it was decided it should be reduced to five years in prison.

The relevant changes in the legislation were not initiated until British parliament recently addressed the issue of the culpability of zoophilic actions. In July 2000, the committee in charge of criminal law revision recommended the penalty for zoophilic interactions be decreased to five years in prison, as a life sentence was considered too drastic and outdated. It was also discussed whether zoophilia was in fact a criminal behavior, and if so, whether it should continue to be observed as a sexual offense under criminal law or be integrated into animal welfare legislation. Contrary to their German colleagues, the British criminal law experts confirmed both the criminality of such behavior, as well as the necessary punishment of it as a sexual offense (Beetz 2002). In particular, this conclusion was justified by concerns that not only human dignity, but also the dignity of the animal was infringed upon by zoophilic interactions, and that an animal could not freely consent to them. Furthermore, it was reasoned that zoophilia was not just an expression of loneliness and closeness, but primarily a sexual crime that reflected deranged behavior. The close connection between animal abuse and sexual crimes was stressed, as well as the existing connection between abuse of animals and children.

The change in legislation was resolved in November 2003, and the new Sexual Offences Act came into force in May 2004. Paragraph 69 is entitled "Intercourse with an animal" and declares both the deliberate and negligent anal or vaginal penetration of an animal with a human penis, as well as the causing and permitting of such action, as punishable crimes. The penalty for this offense is a fine or imprisonment for a maximum of 6 months, or both, if conviction is by a single judge, or a prison sentence of up to two years, in cases of conviction by a jury.

United States and Canada

The North American legislation is influenced in many parts by British law (Miletski 2002). However, in the US there is no national norm about the admissibility of sexual acts with animals in force, so that its regulation is the responsibility of each state. With the exception of Illinois (Miletski 2002) and New Hampshire, zoophilia was considered a severe crime in all the North American states well into the last century. In California, Colorado, Idaho, Missouri, Montana, Nevada, New Mexico and South Carolina it was even prohibited under penalty of life in prison (Christy 1970; Massen 1994). In general, bestiality was subject to the same retributions as homosexuality (Kinsey, Baacke and Hasenclever 1963), of which only very few trials are documented.

The culpability of zoophilia has been abolished in over 20 North American states, with partial reference to the Kinsey Reports, according to which sexual contacts with animals are said to occur mainly during adolescence for a short "experimental phase" (Beetz 2002). In approximately half of the states[2], though, sexual contacts between people and animals are still illegal, where such an offense is considered either a felony or a misdemeanor. While legislation in California, Delaware, Georgia, Minnesota, North Dakota, Utah and Wisconsin explicitly names these offenses as "bestiality" or "deviate sexual act" (California Penal Code 286.5; Delaware Statute Title 11 777; Georgia Statute 16-6-6; Minnesota Statute 609.294; North Dakota Statute 12.1-20-12; Utah Code, 76-9-301.8; Wisconsin Statute 944.17; Idaho Code, Art. 18-6605; Montana Code, Art. 45-5-505; Massachusetts Statute 272 Art. 34; Michigan Penal Code 750.185; Rhode Island Statute 11-10-1; Washington DC Code, Art. 22-3502.), it falls into the general category of "crimes against nature" in many other states. In Idaho, for example, "infamous crime against nature, committed with mankind or with any animal" incurs a prison sentence of at least five years, while in Montana, offenders may be given a prison sentence of up to ten years, or a fine of up to 50,000 US dollars. Other penalties include prison sentences of up to 20 years (Massachusetts); up to 15 years (Michigan); between seven and 20 years (Rhode Island); or imprisonment of up to ten years and/or a fine of up to 1000 US dollars (Washington DC). Contrary to countries of Germanic or Romanic legislations, zoophilia can be used as a reason for divorce in many North American states by civil law (even in some where the offense is exempt from punishment). Some states are discussing the reintroduction of the culpability of this offense.

Apart from the above, there are many federal laws which prohibit zoophilia in the context of other actions in the entire United States. This applies when youths under the age of 18 are involved, or for cases where animal pornography is involved (United States Codes Chapter 71—"Obscenity"). Furthermore, sexual acts with animals owned by a third party are treated as an offense: damage to property.

The legal position in Canada is comparable to the one in the US, but the Canadian criminal code provides a general statement on zoophilia. According to Art. 160, any person committing such an action, or forces another person to such action or entices a person under the age of 14 to such action, will be judged in a summary proceeding and can be penalized with a prison sentence of up to ten years (Criminal Code of Canada, Section 160).

Critical Appreciation

Legal Protection of Sexually Exploited Animals

The analysis on hand not only shows that sexual practices with animals are much more common than generally assumed, but also that such actions are not penalized in most countries by the State any more. The question whether zoophilia—apart from being an ethically debatable issue—is an infringement of the law, has been increasingly answered in the negative since the age of Enlightenment. Only Anglo-American legislation has adhered to the once ubiquitous sanctions, which were first lifted in countries of Romanic and later of Germanic legislations. However, it is generally accepted that zoophilic interactions are not acceptable when the animal involved suffers pain, injury or distress. Most animal welfare legislations therefore penalize such cruelty to animals with quite severe sentences.

So, at least sadistic or violent zoophilic actions are recognized as issues to be regulated by animal welfare legislation which is just and equitable. However, one should question the fact that apart from English-speaking countries, any sexual action involving animals is exempt from punishment as long as there is no proof of cruelty. The lack of specific bans on zoophilia is generally defended with the argument that animals are sufficiently protected by existing legislation. A closer look at the existing regulations shows that this opinion cannot be followed for the following reasons:

1. As laid out before, the national animal welfare laws only protect animals involved in zoophilic actions if it can be proven that they suffered significant pain, injury or distress. However, such acts remain exempt from punishment if the animal does not suffer any of the aforementioned problems, or if the pain or injuries are of a minor nature. The fact that animals are not sufficiently protected in this manner is documented in several surveys. According to a study in Germany at the end of the 1960s, approximately 70% of all zoophilic acts were carried out in a violent manner (often zoosadistically) (Weidner 1972). While these offenses would still be penalized under the existing national animal welfare legislations, the remaining 30% of used animals remain unprotected today because they do not experience any considerable harm (Stettner 1990). It can also be assumed that because there are a large number of unrecorded cases of zoophilia, the percentage of cases that are not considered violent, and hence are exempt from punishment, is much higher. Also, if one is to believe the affirmations of zoophiles, whereby sexual acts with animals generally happen free of any violence and that zoosadistic practices constitute the rare exception (Hoffmann 2003), the number of animals defenselessly exposed to sexual actions is considerably higher than figures would suggest.

Overall, it can be assumed today that zoophilic interactions rarely happen for zoosadistic reasons, although sexual components have to be taken into consideration with any animal abuse cases (Massen 1994). A look into the national statistics of criminal convictions in Switzerland, for example, shows that only one person was convicted of cruelty to animals in the context of a sexual act (see: www.tierimrecht.org/de/faelle). The social taboo on this issue sometimes seems to affect even the investigating authority and courts: in practice, they seem to do their best to avoid investigating or naming the motives behind an obvious act of zoophilia. Also, in practice, the prosecution of cruelty to animals is difficult, as the necessary evidence is not easily provided. In order for an abuse to be penalized according to Art. 27 of the Swiss animal welfare legislation, an animal has to have demonstrably suffered from continuous and agonizing pain. And proving that an animal died as a result of excessive strain due to sexual acts also regularly poses substantial problems to the investigating criminal authorities, especially if the cause of death cannot clearly be identified or an association between the death of animal and the zoophilic act cannot be demonstrated. If there are no veterinary findings of injuries in the genital area, no evidence of other physical injuries related to the incident, and if it cannot be established without a doubt that an animal was stressed, the investigation is dismissed and the accused person is acquitted. Other aggravating factors for zoophilic prosecutions include that the victim is mute and can normally be legally killed—thus, important evidence can be destroyed. If the animal is the property of the accused, and the killing happens free of fear and pain, then this person cannot be held accountable, as most national animal welfare legislations, with the exception of the German and recent Austrian legislation (Ort and Reckewell 2002), do not foresee a general conservation of life for animals (Goetschel and Bolliger 2003).

2. Although the conviction of a person for damage of property can be based on minor damages, this regulation does not provide sufficient legal protection for affected animals. As is the case with the regulation on trespassing, the clauses are not aimed at the well-being and integrity of the animal, but merely at the animal as property of the aggrieved owner. Important to note here is that the law refers to the wounding or killing of another's animal by a third party. If its owner wounds the animal, this is not considered damage of property in the legal sense. In this case, only the (very restricted) national animal welfare legislation can apply. Furthermore, the prosecution for damage of property is only initiated when the aggrieved party lodges a complaint. Third parties are not authorized to lodge such a complaint, so that no enquiry can be initiated without the animal owner's consent. Punishment resulting from damage to property (injury or killing of an animal according to a country's criminal code) is also ruled out if the action was committed negligently. The same applies to acts committed by a third party, who had the owner's permission: the animal was made available for zoophilic acts free of charge or against payment, or the owner got somebody in to engage in zoophilic interactions in order to get sexually aroused. An owner consenting to injury to his animal by a third party is again limited by the law on the prohibition of cruelty to animals: pain and distress must be minimal. Severe cruelty cannot be consented to by owners.

3. Finally, animals are not completely protected from zoophilic abuse by the many existing laws on hardcore pornography, as these regulations are predominantly supposed to protect the sexual integrity of mankind. The public and private presentation of animal pornography, as well as the producing, importing, offering and distributing of the products is generally prohibited, but not the zoophilic action itself. Indeed, whoever participates in such practices can do as they like, as long as they stay within the boundaries defined by the laws of cruelty to animals and damage of property. A conflict with the law only ensues if such persons subsequently describe or document their experiences to third parties in any way; this is an absurd legal position (Massen 1994). When judging hardcore pornography cases, the Swiss federal court regularly states that such cases concern the "depiction of severe perversions, respectively, especially abnormal and repulsive sexual practices" (Schwaibold and Meng 2003). One cannot but agree with this and would suggest that any zoophilic interaction be made illegal.

Violation of the Dignity of the Animal

Modern criminal law concepts are based on the essentially correct tenet that legal provisions should not enforce public morals in an enlightened, liberal and secularized constitutional state, and, in addition, that every penal norm imperatively requires an acknowledged, legally protected right. When determining those legally protected rights today, which are fundamentally violated by any sexual act with animals, historical lines of argument are no longer suitable (Muth 1969). Thus, according to modern interpretations of law, zoophilia cannot be penalized because it offends Divine creation, Christian ethics or public morals or because a person practicing zoophilia hurts society. Already established, legally protected rights, such as those pertaining to morality, human dignity, and the protection of youth, marriage and family, do not justify the protection of animals, and incidentally were not aimed at such a purpose in the past. Rather, they were always subject to predominantly anthropocentric views, although they did at least indirectly protect the animals by leading to general bans on zoophilia.

Today, sexual contact with animals has to be viewed taking into consideration another important factor; one which does not exclusively center on mankind and its interests; one which has remained largely unconsidered by jurisprudence, so far. This factor is the "dignity of the animal," which since 1992 has been protected in Switzerland by the Swiss constitution as one aspect of the "dignity of the creature" (Teutsch 1995; Krepper 1998; Baranzke 2002; Goetschel 2002; Goetschel and Bolliger 2003). To date, Switzerland is the only country in the world to have implemented this factor into its constitution.

Modern animal welfare legislations are based on the tenet that animals are emotional creatures and thus are to be respected and protected, not only in the interest of mankind but also for their own sake. The recognition of the dignity of animals is one of the main pillars in modern animal welfare concepts, which in law has led to animals being seen as more than mere objects or things. This idea transcends mere prevention of suffering, pain, injury and fear, and denotes a general respect for the physical and mental integrity of every individual animal. This includes, for example, protection from humiliation, excessive exploitation and interference with an animal's appearance, as well as the restriction of certain kinds of contacts with animals which are not linked to obvious injury, but which concern other animal interests and are to be respected by mankind.

According to this view, one important aspect of the dignity of the animal is its sexual integrity. Along the lines of the sexual integrity of mankind—which is recognized in modern criminal law concepts as a legally protected right—it includes unhindered sexual development and sensation, the protection from damaging decision-making by sexual exploitation of dependencies, and the protection from sexual harassment. The dignity of animals is thus not only injured by violent sexual acts, but any zoophilic act which does not respect the intentions of the animal, and therefore is effected by using some form of force. This is especially important, as not all animals suffer physical injury or can show clearly their aversion to intimate contacts with people. Indeed, there are many animals who remain relatively unexpressive during these sexual acts and who obediently submit to them. If there is psychological stress going on here, it is not easy for us to detect.

That zoophilic relationships can be mutual and that animals can develop strong affections for people, including a sexual component, is not disputed (Rosenbauer 1997). It does not appear to be difficult for some animals to enter into an intimate relationship with a person, and it can be quite easy to sexually arouse and satisfy a male animal. Indeed, sometimes animals (allegedly) voluntarily participate in zoophilic acts or even take the initiative (Massen 1994; Muth 1969; Dekkers 1996). However, in general, an animal only does this if it is used to such behavior, that is, it has been trained to perform this behavior and has thus become focused on a human sexual partner (Frey 2003). As a result of such training, these animals offer little or no resistance when sexually approached by a person (Frey 2003). Apart from this trained behavior, some natural reflexes and instinctive acts are exploited for zoophilic contacts (Muth 1969; Weidner 1972; Stettner 1990). Such conditioning does not only infringe upon the free sexual development of an animal, but also holds the danger of the creation of a strong dependency. Thus, an animal that has repeated sexual intercourse with a person can commit to this person to such a degree that it loses all interest in sexual interactions with others of its own kind (Ford and Beach 1954; Masters 1965).

The violation of an animal's sexual integrity, thus firstly does not depend upon the question of what an animal feels during a zoophilic act, but rather whether such an act complies with its free will. People generally cannot discern whether such a behavior by an animal happens voluntarily. Because of the communication barrier between man and animal, it naturally remains unclear what exactly an animal feels during a zoophilic interaction if it does not show any evidence of pain, suffering or injury. As is the case with humans, it has to be assumed that any damage to the well-being of animals can only partially be determined from subsequent clinical investigations (Luy 2003). One cannot even say whether those animals who were sexually imprinted to human beings feel good during sexual intercourse with a person.

Whether zoophilia in fact ever happens with mutual consent, that is, is wanted and appreciated by the animal, can only be speculated on. Rather, one should act on the assumption that the animal's consent is forced, either through an artificial fixation on a person or by use of other psychological methods. The labeling of such acts as "animal love" or "sexuality in partnership," as the people involved often call them, misjudges such circumstances and seem euphemistic in the light of the different methods in practice (Hunold 1970; Massen 1994).

Despite the continuing affirmations of zoophiles, that their intimate relationships with animals are not characterized by violence and subordination, but rather by a mutual attitude of respect and trust, it remains a fact that animals are first and foremost exploited to satisfy the sexual urges of people, and are degraded as sexual objects, even if the intercourse remains free of violence (Goetschel, Bolliger and Gerritsen 2005). Admittedly, in our society many animals are used against their will for other purposes, such as animal testing or the production of food. The acceptability of these uses is called into question if one applies the principle of the "dignity of the animal." However, different to zoophilia, most of these actions can be socially justified, as long as they abide by the legal conditions set down.

Other arguments from supporters of zoophilia fail when one considers the violation of the dignity of an animal. For example, it is completely irrelevant that no offspring can result from zoophilic interactions and that usually the physical health of man and animal is unharmed if precautionary measures are taken (Muth 1969). Also, the comparison with homosexuality, which is often made by zoophiles is not appropriate.

Conclusions

This paper demonstrates that sexual acts with animals are not just of peripheral importance in our society; but they are turned into peripheral matters by making such acts publicly and legally taboo. The vast amount of related material on the Internet, which invites people to engage in such actions and which makes one believe that zoophilia is nothing but a harmless variety of sexuality—a new lifestyle—disproves that it has minor social impact.

In the 1950s, the Kinsey research showed that the frequency of sexual acts with animals was at a high level and that such actions would occur more frequently if the conditions to commit them were more convenient (Kinsey et al. 1955). This is exactly what has happened: today, with the exception of a few Anglo-American countries, most states only prosecute such deeds if substantial and proven injury or pain is inflicted on an animal. Zoophilia represents a problem in animal welfare *in general*, regardless of how widespread it is and the degree of physical violence involved (Stettner 1990). As the social aversion towards sexual abuse of animals by mankind obviously does not sufficiently restrain zoophilia, the problem has to be solved by national legislators.

Over the past few decades, ethical animal welfare concepts which impose the responsibility for the care of an animal on its owner, have been gaining in importance, both nationally and internationally. Against this backdrop, zoophilic interactions have to be considered punishable, not only in cases of obvious abuse, but also on the basis of possible psychological damage incurred by the animal during the act (Frey 2003; Buschmann 2004). Because zoophilia infringes unquestionably upon the sexual integrity of an animal, it above all represents a violation of the dignity of an animal, and thus constitutes a fundamental infringement of an animal's welfare (Goetschel 2002).

Apart from a few Anglo-American countries/states, animals are not provided with sufficient legal protection from sexual exploitation. Far from it. The legal requirement for proof of obvious injuries creates a loophole for exploitation, and denies animals a right against despotic sexual acts. Indeed, the opinion that generally everything that happens by mutual consent between the parties involved is allowed in sexual interactions, and that the state should not dictate nor prohibit certain practices, has been accepted in most cultures. But this kind of tolerance must not be abused at the expense of the animal, and the line should be drawn where a sexual partner, with equal rights, does not consent to an action. The fact that animals, because of being under human care, are not considered equal in this sense, means their consent to zoophilic interactions can only be assumed.

Many national legislators have repeatedly and explicitly declared their support for the obligation to provide for the welfare of an animal in custody. With regard to sexual interactions, however, such obligations are not formulated in such a way that the risk of physical or mental injury is minimized from the outset. Thus, the legal distinction between admissible and inadmissible zoophilia neither takes into consideration that an animal's suffering can only rarely be evidenced, nor that the animals affected are creatures in need of protection, ones who cannot stand up for their own rights and who are abused mostly covertly. In order to "nip zoophilia in the bud" and to avoid development of other types of abuse, clear and general bans on zoophilia are necessary (Luy 2003). Even a constitutional protection of the dignity of the animal (which presently exists only in Switzerland) is not enough, although the basic principle could be directly put into use. According to the penal law maxim "no punishment without law," a specific criminal norm is required that defines in detail which behaviors are punishable and which are not.

Offenses could be entitled "sexual acts with animals" for example, and the wording could follow the proscriptions of human sexual acts with children (Bolliger 2000; Goetschel 2002). Despite that the doctrine still rejects such a comparison by the majority (Schwaibold and Meng 2003), the unquestionable fact that animals can feel physical and psychological pain (Luy 2003) surely means they should be afforded a similarly restrictive legal protection of sexual exploitation, as are children and other people. (Luy 2003). It is generally recognized that the sexual freedom of an individual ends where the right for self-determination of another begins. People who cannot assert their legal positions on their own, (such as children and people with disabilities), are rightfully protected by modern laws through restrictive acts. Sexual contacts with these persons are prohibited because of their basic need for protection, and it is irrelevant whether they possibly participated voluntarily in such acts or were physically injured by them. To put animals into this category with people, in respect of their being worthy of protection from sexual exploitation, is only a consistent development both ethically and legally, and does not constitute an improper humanization of animals.

For reasons of orderly legal considerations, general bans on zoophilia should not be included in the national penal code, but rather should be added as an offense to the catalogue of forbidden actions in animal welfare legislations (Schwaibold and Meng 2003), as is currently being stipulated by animal welfare organizations in the ongoing revision of the Swiss animal welfare legislation. Moreover, physical injuries of animals as a consequence of zoophilic interactions should remain punishable as an offense of animal abuse. The only exception to any sexual act with an animal should be veterinary interactions, for example, assisting the reproduction of animals.

The question of exactly which acts fall under the term "sexual acts with an animal" can generally be based on the jurisdiction of the national laws on animal pornography. Of course not every single physical contact is to be considered a zoophilic action. The touching of human genitalia by the muzzle of an animal, for example, is not necessarily an animal welfare problem *per se* and often is only an instinctive behavior. However, if a person systematically exploits such behavior patterns, then the actions cannot be reconciled with the dignity of the animal. Sexual intercourse between a person and an animal, of either the same or opposite sex, as well as any action resembling sexual relations (i.e., with the deliberate intention to achieve sexual satisfaction by physical contact of the human's or animal's sex organs with the human or animal body; Merki 1948), and regardless of whether an orgasm was reached by the person or animal involved, would undoubtedly be punishable.

Apart from zoophilia itself, all actions related to it, such as the training and attuning of animals to perform the act, and the placing, relinquishing and making available of animals for sexual purposes should be prohibited. Also to be outlawed are all actions related to animal pornography—not only the manufacturing and distribution, but also, as is the case in the Swiss legislation, the acquisition and ownership of the products (Schwaibold and Meng 2003). This should be done, as zoophilic depictions or demonstrations could have a motivating effect on the consumer, increasing his or her willingness to imitate such actions (Schwaibold and Meng 2003). And as a close connection between violent assaults on animals and on humans has been established (Hunold 1970; Illi 1998), and that there is the possibility of lasting damage to the sexual development of children and adolescents, attention should be especially focused on the containment of the vast amount of zoophilic material on the Internet.

In conclusion, it can be said that zoophilia is an extremely complex and interdisciplinary issue, which is not only subject to a whole array of esthetical, historical, ethical and religious taboos, but is also relevant to various scientific fields, such as anthrozoology, psychology, sociology, humanities and veterinary medicine. But, above all, it remains a problem of animal welfare. In recognizing animals as our sentient fellow creatures, we should afford them the same respect for life, dignity and physical integrity as we would humans. This can only be achieved effectively through the law. The general model for ethical animal welfare has substantially changed since the time of the abolition of the general national bans on zoophilia, such that the reintroduction of these laws is now conceivable. Only by doing so can this issue, which is traditionally clouded by a certain "fog of repulsion, outrage and sensation" (Merki 1948), be effectively conceived, and the exploitation of animals for sexual purposes can be brought to an end. This would lead, ultimately, to an intensification of relationships with animals, based on actual partnership and not on exploitation.

Acknowledgements

The authors thank Martina Leuthold, Lena Hildermann, Oliver Goetschel and Susan Platt-Läubli for their valuable assistance with the research, as well as their help with editing the article.

Notes

1. A penalty based on the income and assets of the convicted person.
2. Arkansas, Delaware, Georgia, Idaho, Indiana, California, Kansas, Louisiana, Maryland, Massachusetts, Michigan, Minnesota, Mississippi, Montana, Nebraska, New York, North Carolina, North Dakota, Oklahoma, Rhode Island, South Carolina, Tennessee, Texas, Utah, Virginia, Washington DC and Wisconsin (Miletski 2002).

References

Arnold W., Eysenck H. J. and Meili R. eds. 1997. *Lexikon der Psychologie.* Augsburg: Bechtermünz Verlag.

Baranzke, H. 2002. *Würde der Kreatur? Die Idee der Würde im Horizont der Bioethik.* Würzburg: Königshausen und Neumann.

Beetz, A. 2002. *Love, Violence and Sexuality in Relationships between Humans and Animals.* Doctoral dissertation. Aachen, Germany: Shaker Verlag.

Berg, S. 1963. *Das Sexual-Verbrechen.* Hamburg: Kriminalistik.

Bolliger, G. 2000. *Europäisches Tierschutzrecht.* Doctoral dissertation. Zürich: Schulthess Juristische Medien AG.

Buschmann, J.-U. 2004. *Anregungen zur Berücksichtigung psychischer Beeinträchtigung als Schaden im Sinne des § 17 TierSchG—Notwendigkeit eines etho/physiologischen Nachweises traumatisierender Einflüsse.* Bochum: www.verschwiegenes-tierleid-online.de/essay_j.u.-buschmann.htm

Christy, R. 1970. *Sodomie—Die Zunahme einer sexuellen Abartigkeit nach privaten und amtlichen Quellen.* München: Lichtenberg Verlag.

Dekkers, M. 1996. *Geliebtes Tier—Die Geschichte einer innigen Beziehung.* Reinbek: Hanser Carl Verlag.

Ford, C. S. and Beach F. A. 1954. *Das Sexualverhalten von Mensch und Tier.* Berlin: Colloquium Verlag Otto H. Hes.

Friday, N. 2004. *Die sexuellen Phantasien der Frauen.* 23. Aufl. Reinbek: Rowohlt Verlag GmbH.

Frey, G. nd. *Sodomie: Der sexuelle Missbrauch von Tieren.* www.dogs4life.de/printable/tierschutz/sodomie/index.html.

Frey, G. 2003. *Sexuelle Handlungen mit Tieren: Die Rechtslage in Deutschland—ein Überblick,* Lentföhrden: www.verschwiegenes-tierleidonline.de/rechtslage.teil%201.htm

Goetschel, A. F. 1986. *Kommentar zum Eidgenössischen Tierschutzgesetz.* Bern/Stuttgart: Haupt Paul Verlag.

Goetschel, A. F. 2002. Würde der Kreatur als Rechtsbegriff und rechtspolitische Postulate daraus. In *Die Würde des Tieres,* 141–180, ed. M. Liechti. Erlangen: Harald Fischer Verlag.

Goetschel, A. F. and Bolliger, G. 2003. *Das Tier im Recht—99 Facetten der Mensch-Tier-Beziehung von A bis* Z. Zürich: Orell Füssli Verlag.

Goetschel, A. F., Bolliger, G. and Gerritsen, V. 2005. Argumentarium Zoophilie, Zurich: www.tierimrecht.org/de/argumentarium/zoophilie.php.

Grassberger, R. 1968. *Die Unzucht mit Tieren.* Wien/New York: Springer.

Guggenbühl, D. 2002. *Mit Tieren und Teufeln—Sodomiten und Hexen unter Basler Jurisdiktion in Stadt und Land 1399 bis 1799.* Basel: Verlag des Kantons Basel.

Hoffmann, A. 2003. *Das Lexikon der Tabubrüche.* Berlin: Schwarzkopf und Schwarzkopf.

Hunold, G. 1970. *Abarten des Sexualverhaltens: Ungewöhnliche Erscheinungsformen des Trieblebens.* München: Wilhelm Heyne Verlag.

Illi, T. 1998. Sodomie: Der Frevel an den Tieren. *Beobachter* 5: 18–23.

Kinsey, A., Baacke, M. and Hasenclever, W. 1963. *Das sexuelle Verhalten der Frau.* Berlin/Frankfurt: S. Fischer Verlag.

Kinsey, A., Pomeroy, W. B., Martin, C. E., von Eckardt-Jaffe, M. and Seemann, W. 1955. *Das sexuelle Verhalten des Mannes.* Berlin/Frankfurt: Fischer Verlag.

Krepper, P. 1998. Zur Würde der Kreatur in Gentechnik und Recht. Dissertation. Basel/Frankfurt: Helbing und Lichtenhahn.

Krings, S. 1995. Sodomie am Bodensee: Vom gesellschaftlichen Umgang mit sexueller Abartigkeit in spätem Mittelalter und früher Neuzeit auf St. Galler Quellengrundlage. In *Schriften des Vereins für Geschichte des Bodensees und seiner Umgebung*, 113, 1–45. Friedrichshafen: Author.

Luy, J. 2003. *Sodomie—Tierschutzethische und rechtsphilosophische Aspekte.* Berlin: www.verschwiegenes-tierleid-online.de/essays.luy

Massen, J. 1994. *Zoophilie—Die sexuelle Liebe zu Tieren.* Köln: Pinto Press.

Masters, R. E. L. 1965. *Sexuelle Tabus und Moral, Eine wissenschaftliche Untersuchung perverser sexueller Praktiken in verschiedenen Kulturen.* Hamburg: Kala Verlag Krohn KG.

Merki, P. 1948. Die strafrechtliche Behandlung der Unzucht mit Tieren (besonders in der Schweiz). Dissertation. Schwyz: Author.

Miletski, H. 2002. *Understanding Bestiality and Zoophilia.* 2nd edn. Bethesda, MD: Author.

Muth, M. F.-W. 1969. Zur Frage der Berechtigung einer Strafnorm gegen die Unzucht mit Tieren. Dissertation. Universität Freiburg im Breisgau.

Ort, J.-D. and Reckewell, K. 2002. Kommentar zu §§ 17-20a TierSchG. In *Tierschutzgesetz: Kommentar,* 135–163, ed. L.H.G. Kluge. Stuttgart: Kohlhammer Verlag.

Rosenbauer, F. 1997. Sexueller Kontakt mit Tieren. Siegen. www.rosenbauer.de/ha zo.htm

Schwaibold, M. and Meng, K. 2003. Kommentar zu Art. 197 StGB. In *Strafgesetzbuch, Basler Kommentar*, 1062–1076, ed. M. A. Niggli and H. Wiprächtiger. Basel/Genf/München.

Stettner M. 1990. Unzucht mit Tieren - ein Tierschutzproblem. *Deutsche tierärztliche Wochenschrift* 97: 171–174.

Teutsch, G. M. 1995. *Die Würde der Kreatur.* Bern/Stuttgart/Wien: Haupt Paul Verlag.

Vogel, U. 1980. Der bundesstrafrechtliche Tierschutz. Dissertation. Zürich: Schulthess Juristische Medien.

von Hentig, H. 1962. Soziologie der zoophilen Neigung. Stuttgart: Ferdinand Enke.

Weidner, E. 1972. Sodomie und Sadismus als Tierschutzproblem. Dissertation. University of Giessen.

Bestiality and zoophilia: Associations with violence and sex offending

Andrea M. Beetz

Department of Veterinary Medicine, University of Cambridge, UK

Abstract

Bestiality is a subject that usually evokes strong emotional reactions, such as abhorrence, disgust, and ridicule. During the past few decades it has gained more interest from animal protection workers and researchers in the field of interpersonal violence and sex offending. This is not surprising, as a growing body of research indicates a link between animal abuse and interpersonal violence. This article provides an overview of data on bestiality regarding its prevalence in normal populations in comparison to samples of special populations, with a focus on violence and sexual offenses. In addition, the importance of differentiating between the various forms of bestiality (depending on the species of animal involved and the different sexual practices engaged in) and the levels of violence used are addressed, as will be the question of consent: what is its role in the perception and evaluation of bestiality, and the more complex phenomenon of zoophilia? Based on the accumulated information, the relevance of bestiality to the fields of interpersonal violence, sex offending, law, and animal protection is discussed.

Keywords: *animal abuse, bestiality, interpersonal violence, sex offending, zoophilia*

Sexual contact with animals, usually called bestiality, is a phenomenon rarely mentioned in either casual or scientific communications. In a non-scientific context, the subject of bestiality is often addressed in jokes or is viewed as one of the most bizarre behaviors, and it provokes strong emotional reactions, such as disbelief, disgust, and abhorrence. The general public regards people, mainly men, who engage in bestiality as mentally disturbed, perverted, dangerous, and capable of other violent and sexual offenses. Furthermore, parallels to sexual abuse of children are frequently drawn. However, few scientific studies have been conducted on bestiality; a small number of studies have focused directly on this sexual practice, while

Address for correspondence: Dr A. M. Beetz, Schillerstrasse 28a, 91054 Erlangen, Germany. Ph: +49 09131-4000 455; e-mail: andrea.m.beetz@gmx.de.

studies on the broader subjects of interpersonal violence and sexual abuse have sometimes included bestiality as one of many behaviors assessed. Results from these broader studies are usually cited to support the claim of a connection between bestiality and interpersonal, in particular sexual, violence. This accords with another established link between animal abuse in general and interpersonal violence (for an overview, see Lockwood and Ascione 1998). However, new insights from recent studies, especially the ones conducted by Beetz (2002), Miletski (2002), and Williams and Weinberg (2003), which focus exclusively on bestiality and zoophilia, show the necessity for a more differentiated discussion and further investigation. In addition, differences in the legal status of bestiality, depending on the country or state, complicate matters further. From the viewpoint of animal protection agencies, the question of whether animals can give consent to sexual acts being carried out on them also needs to be considered.

This paper provides an overview of the existing information on bestiality and zoophilia, to allow an objective and more differentiated discussion of a possible link between sexual contacts with animals, animal abuse, and interpersonal violence. In the next section, information on the prevalence of bestiality in normal and special populations will be reported. This information should serve as a reference point for an evaluation of the prevalence of this behavior among samples of violent and sex offenders described later on. Then, an overview of the findings of recent research, in particular the studies of Beetz (2002), Miletski (2002), and Williams and Weinberg (2003), will be given with a focus on the different types of sexual contact people have with animals, the various sexual practices, and the kinds of animals involved. This leads on to a discussion of the necessity to assess the degree of violence involved in sexual practices with animals when discussing a link with interpersonal violence. Following this, a comparison is made between bestiality and other procedures performed on animals, and special attention will be given to the question of consent and the legal status of bestiality in different countries. Finally, I will integrate the information available and make some conclusions about the link between sexual contact with animals and interpersonal violence and sex offending.

The Prevalence of Bestiality

As a basis for comparison with the prevalence rates found among offenders reported in the next section, some data on the prevalence of sexual contacts with animals are reviewed here. Few studies provide such information for a normal population; the most important ones are those by Kinsey, Pomeroy and Martin (1948), Kinsey et al. (1953), and Hunt (1974).

The famous Kinsey reports on the sexual behavior of the American male and female also include data on sexual contacts with animals. Kinsey and his colleagues interviewed about 5300 adult, white men and 5800 adult, white women about their sexual experiences. Although the objectivity of the methods employed is sometimes criticized, the studies provide important information. Kinsey and his colleagues found that in rural areas about 40 to 50% of the males had had at least one sexual encounter with an animal, and 17% had even experienced an orgasm as a result of sexual contact with animals during adolescence. Amongst all the American men in the study, however, the prevalence was much lower, about 8% (Kinsey, Pomeroy and Martin 1948)—still a substantial number, though. With such a sensitive subject as bestiality, behaviors were probably under- rather than over-reported.

For the female population, Kinsey et al. (1953) found a prevalence rate of 3% for sexual contacts with animals. The frequency of the contacts ranged from just once to several times a year to several times a week, and sometimes these contacts occurred over a period of several years or even a whole lifetime. These contacts included oral practices and masturbating (by rubbing against the body of the animal), as well as vaginal and anal penetration. It needs to be mentioned that the number of persons who repeatedly had sexual contact with animals over a longer period of time was much lower. Many participants indicated that contacts were motivated by curiosity or occurred due to particular circumstances.

About 20 years later, Hunt (1974) reported data on 982 American men and 1044 American women, including information on bestiality. Even though the prevalence rates were lower than those reported by Kinsey, Pomeroy and Martin (1948) and Kinsey et al (1953)—about 5% of the men and about 2% of the women reported at least one sexual encounter with an animal—Hunt (1974) suggested that this was due to urbanization, with people having less access to animals in cities. Hunt (1974) also found that the majority of men (75%) and women (67%) who had had sexual contacts with animals reported it had occurred on only a few occasions. As in Kinsey's studies, Hunt found that the majority of sexual contacts with animals occurred during adolescence or early adulthood.

Alvarez and Freinhar (1991) provided information on the occurrence of sexual contacts with, and sexual fantasies about, animals among a special sample: 20 psychiatric in-patients, 20 medical in-patients, and 20 psychiatric staff members. They found that 55% of the psychiatric in-patients, but only 10% of the medical in-patients and 15% of the psychiatric staff members, had had at least one sexual fantasy or contact with an animal. Unfortunately, the authors did not distinguish between actual contact and fantasy. However, based on the prevalence rates found, they concluded that questions regarding such fantasies or contacts should be routinely included in a psychiatric assessment.

Even if it does not directly relate to the frequency of actual sexual contacts with animals, it might be of interest to know more about the prevalence of sexual fantasies about animals among the normal population. According to Kinsey, Pomeroy and Martin (1948) and Kinsey et. al. (1953), 1 to 2% of American men and 0.6 to 1.3% of American women fantasized about having sex with animals. In a more recent study, with a smaller sample of 180 mainly heterosexual German men and women, Schweitzer and Beetz (2000, cited in Beetz 2002) found that 8.3% (9.3% of the men and 7.4% of the women) *seldom* fantasized about getting sexually aroused by an animal, while 4.4% (3.5% of the men and 5.3% of the women) *sometimes* had sexual fantasies about animals. The desire to experience their fantasy in reality was rated as medium to very strong among 1.7% of the sample. And yet, interestingly, 4.4% of the sample (4.7% of the men and 4.3% of the women) reported actual experiences of sexual arousal by an animal (Schweitzer and Beetz 2000, cited in Beetz 2002).

Overall, it can be seen from the available data that sexual contact with animals occurs at a not infrequent rate in Western cultures, and thus deserves further investigation.

Bestiality and Animal Abuse in the Context of Interpersonal Violence and Sex Offending

As stated earlier, data on bestiality in connection with interpersonal violence, and in particular sex offending, are usually obtained from studies where sexual contact with an animal was not the primary focus, but was included as one of many behavioral markers. Most often in these studies only the issue of animal abuse is assessed. And as a growing amount of research now exists documenting a link between animal abuse and interpersonal violence (for an overview see Lockwood and Ascione 1998; Ascione and Arkow 1999), it is important to reconsider where bestiality fits in. In general, bestiality is automatically considered to represent animal abuse, even though definitions of bestiality vary widely between studies, from oral or masturbatory contact to sexual penetration.

Conduct Disorder, Antisocial Personality Disorder, and Animal Abuse

Animal abuse has been recognized as one of the early warning signs of a psychological dysfunction, in particular conduct disorder in childhood and adolescence. Indeed, animal abuse is listed as one of the symptoms for this disorder in the *Diagnostic and Statistical Manual of Mental Disorders* (DSM-IV), published by the American Psychiatric Association (APA 1994). Conduct disorder is characterized by a "repetitive and persistent pattern of behavior in which the basic rights of others or major age-appropriate societal norms or rules are violated" (APA 1994, p. 85). This includes aggression to people and animals, destruction of property, deceitfulness and theft, and serious violation of rules (APA 1994). The diagnosis of conduct disorder is applied mainly in childhood and adolescence. A significant proportion of juveniles with this diagnosis, however, continue to show these behaviors as adults, allowing for a diagnosis of Antisocial Personality Disorder, also referred to as psychopathy or sociopathy (APA 1994). Both diagnoses are closely related to criminal behavior, and animal abuse has also been linked directly to Antisocial Personality Disorder (Gleyzer, Felthous and Holzer III 2002) as well as adult criminal offending (Merz-Perez, Heide and Silverman 2001).

Animal abuse can occur in very different forms, ranging from active maltreatment, which includes bestiality (Vermeulen and Odendaal 1993), to passive neglect or commercial exploitation. As with child abuse, emotional, physical, and sexual abuse of animals can be distinguished (Ascione et al. 2003). In general, animal abuse can be defined as "socially unacceptable behavior that intentionally causes an animal pain or distress and may result in an animal's death" (Ascione 1993).

Abuse History and Animal Cruelty

Animal cruelty is frequently linked to a history of interpersonal abuse; the experience of corporal punishment in adolescence (Flynn 1999), physical abuse (DeViney, Dickert and Lockwood 1983), and sexual abuse (see below) significantly increases the likelihood of animal abuse occurring. Flynn (1999) studied 267 male undergraduates and found that, of those who experienced corporal punishment in adolescence, 57% reported active animal abuse, compared to only 23% of the men who had not been physically disciplined. In a study by DeViney, Dickert and Lockwood (1983), 60% of the 53 families they assessed—who were known for child abuse and neglect and had pets—also displayed animal abuse or neglect, perpetrated by the parents and the children.

Of special interest here is the potential connection between sexual abuse of a person and bestiality. It seems plausible that if a person experienced a history of sexual abuse, this may lead him/her to sexually or physically abuse animals. Ressler et al. (1986) used official records and interviews to investigate the connection between sexual abuse in childhood or adolescence and sexual activities, interests, and deviations among a sample of 36 sexually oriented murderers. They found that 43% of the participants had been sexually abused in childhood, 32% in adolescence, and 37% as adults. Those sexually abused in childhood were more likely to report active cruelty to animals in childhood (58% versus 15% of those not sexually abused in childhood), cruelty to children (73% versus 38%), and assaulting adults (58% versus 15%). Of the perpetrators who had experienced sexual abuse as adolescents, 67% (versus 31% of the non-abused) engaged in animal abuse. Twenty-three percent of the perpetrators indicated an interest in sexual contact with animals.

In their study of a normative sample of 880 children and 276 sexually abused children, aged 2 to 12 years, Friedrich et al. (1992 cited in Ascione et al. 2003) found significantly higher rates of animal abuse among the sexually abused group; 34.8% of the boys and 27.5% of the girls, compared to 4.9% of the boys and 3.3% of the girls in the normative group. Another study documenting a higher prevalence of animal cruelty among sexually abused children was conducted by McClellan et al. (1995).

Animal Abuse and Interpersonal Violence

A growing number of studies document the existence of a link between animal abuse and interpersonal violence, even though neither causal nor chronological relations between these two phenomena have been established for certain. Hellmann and Blackman (1966) studied 53 male prisoners who had committed non-aggressive crimes and 31 male prisoners convicted of violent crimes. Seventeen percent of the first group reported cruelty towards animals, while significantly more (52%) of the violent criminals had abused animals. In their study of homicidally aggressive children in an inpatient ward, Lewis et al. (1983) found that 14% had a record of animal cruelty, in comparison to only 3% of non-homicidally aggressive children.

Similarly, among a sample studied by Felthous and Kellert (1986), 25% of criminals classified as aggressive, but only 5.8% of the non-aggressive criminals and none of the non-criminals, had abused animals in childhood more than five times. Noteworthy is the distinction the authors made between abusive acts according to the severity, frequency, and the species of animal and its social value. Also, the criminals were classified as either low- or high-aggressive; substantial animal abuse was significantly related to the impulsive and recurrent violence of the high-aggressive group. Felthous and Kellert (1987) emphasized the importance of assessing the severity and frequency of animal abuse when investigating links to interpersonal violence.

Arluke (1998) studied criminal behavior among 153 individuals who had a record of intentional animal cruelty. Seventy percent of these animal abusers also had records for other crimes: related to drugs, property, public disorder or violence. Compared to a matched-control group with no record of animal abuse, the animal abusers were five times more likely to have committed violent crimes towards people (Arluke 1998; Arluke et al. 1999). Schiff, Louw and Ascione (1999) found that among a sample of 117 incarcerated men, 63.3% of those who had committed aggressive crimes reported committing animal cruelty in childhood, compared to only 10.5% of the men who had committed non-aggressive crimes. Overall, the link between aggression towards animals and aggression towards humans seems well established. Variations in the rates of reported animal abuse are very likely caused by different methods of data collection and the frequency and severity of the violent acts.

A high prevalence of animal cruelty amongst a special sub-group of criminals—sex offenders—was documented in the following studies. In their investigation of sexual homicide perpetrators, Ressler et al. (1988, cited in Ascione 1993) found prevalence rates for animal cruelty of 36% in childhood and 46% in adolescence. Among a sample of 27 juvenile, male sex offenders assessed by Beetz (2002), 37% admitted to having abused animals. Tingle et al. (1986) studied 64 male sex offenders and found that 48% of the rapists and 30% of the child-molesters had reported animal cruelty. These numbers show that between a third and one half of the definitely aggressive sex offenders, such as sex murderers and rapists, had abused animals, while child molesters, of whom not necessarily all employ violence in their offenses, show somewhat lower rates.

Sexual Contact with Animals Among Specific Populations

A few studies have compared sexual contacts with animals among different populations, mostly groups with a history of abuse, sex or non-sex offending, or other clinical problems. A recent study by Ascione et al. (2003) provided information on the prevalence of animal cruelty and "touching an animal's sex parts" among 1433 children aged 6 to 12 years, using checklists answered by the maternal caregivers. The sample was divided into three groups: a normative group with 540 children, a group of 481 sexually abused children, and a group of 412 psychiatric outpatients. Overall, cruelty towards animals was reported for 3.1% of the normative group, 17.9% of the sexually abused group, and 15.6% of the psychiatric outpatients. "Touching animals' sex parts" was reported much less frequently: only 0.4% of the normative group and 0.9% of the psychiatric outpatients. However, 6.3% of the sexually abused children had done this. The study also showed that, overall, cruelty to animals was significantly related to cruelty to others and touching an animal's sex parts. Furthermore, touching an animal's sex parts was significantly related to cruelty to others. Differentiating by group, these correlations were found only among the sexually abused children. Ascione et al. (2003) concluded that "sexualized forms of cruelty may be more specifically related to a history of sexual abuse" (Ascione et al. 2003, p. 206).

In a study of 100 violent criminals and a normal, matched-control group (*n* = 75), ten Bensel et al. (1984) compared animal-related experiences, including sexual interactions with animals. None of the violent offenders reported sexual contact with animals and, according to the authors, reactions to that question were very emotional and negative. Among the control group, 3.9% admitted to sexual contact with animals. Miller and Knutson (1997) assessed sexual interactions with animals in their study of 314 prisoners. Twenty-two participants (7%) had watched sexual contact between animals, 16 participants (5.1%) had touched an animal sexually, and nine inmates (2.9%) had had sexual intercourse with an animal. Overall, approximately 11% of this criminal sample had seen or engaged in sexual contacts with animals.

Several studies which include the subject of bestiality can be found in the field of sex offending. As mentioned earlier, Ressler et al. (1986) reported that 23% of their sample of sex murderers indicated an interest in sexual contact with animals. An important early study in this field, investigating the prevalence of bestiality among over 1000 white, male, convicted sex offenders, was conducted by Gebhard et al. (1965). For the purpose of comparison, data were also obtained from a sample of 881 men who had been convicted of felonies or misdemeanors, but never sex offenses (prison group), and a group of 471 white males who had not been convicted of anything more serious than traffic violations (control group). Gebhard et al. (1965) restricted their study to sexual contacts with animals occurring after the onset of puberty, and included penile penetration of the animal. The lowest prevalence rate for animal contacts, 8.3%, was found for the control group and the group of heterosexual sex offenders who abused adults. This rate is comparable to the findings of Kinsey, Pomeroy and Martin (1948) for the general population. Sex offenders who had abused female minors (non-incestuously) reported the highest prevalence rate, 33.3%. Among the other sub-groups of sex offenders (homosexual vs. heterosexual; child, juvenile or adult victims; incestuous, non-incestuous), 12.5% to 24% had engaged in bestiality.

Frazier (1997, cited in Ascione 1999) found high rates (37%) of bestiality and sexual abuse of animals in a sample of 30 sexually violent juvenile offenders. Beetz (2002) reported that only two out of 27 (7.4%) juvenile sex offenders she interviewed admitted to having watched others engage in bestiality, and another two (7.4%) admitted to having engaged in sexual contact with animals themselves. Asked about sexual fantasies involving animals, 59.3% denied such fantasies, while the rest of the group did not answer this item at all. Especially with such a delicate subject as bestiality, the honesty of responses should be questioned, and one can assume that the results under-report the true prevalence rates.

A study that probably provides more accurate numbers in regard to bestiality amongst sex offenders was conducted by English et al. (2003). The authors compared information obtained from offenders after their conviction (when monitored by a polygraph) with the information provided before (when not monitored by a polygraph). Among the 180 convicted sex offenders investigated, prior to the polygraph investigation only 4.4% reported having ever engaged in bestiality, while during the polygraph investigation 36% admitted to having engaged in it.

In a study of 381 institutionalized, male juvenile offenders (Fleming, Jory and Burton 2002), 6% of the participants reported that they had done something sexual to an animal or animals without having been forced to do so, while 42% admitted to offending sexually against other persons. Twenty-three of the 24 juveniles who had reported sexual acts with animals had also sexually offended against humans. Only 12 of them had been prosecuted for sex offenses, indicating a high number of undetected cases and suggesting a strong connection between sexual offenses towards humans and animals. Fourteen of the 24 juveniles had masturbated against the animals, while ten juveniles even indicated penile penetration of the animal. Four juveniles had oral contact with the genitals of animals, six had inserted a finger, and two juveniles had inserted an object into an animal. As Fleming, Jory and Burton (2002) stated, the motivation for these acts could have been sexual curiosity, as well as sexual gratification for the person, sexual sadism, or a combination of these.

Sex with animals has been found to be practiced in connection with masochistic and sadistic practices, or a combination of both. Relations to sadomasochism were addressed in a study by Miletski (2002), who gathered data from a voluntary sample of 93 men who admitted to sexual practices with animals, using an anonymous questionnaire. Only 4% of the men fantasized "always" or "primarily" about sadomasochistic sex. However, 9% indicated that they had "forced someone to do something sexually that the person did not want to do" (Miletski 2002, p. 118). In a sample of 113 men who had engaged in sexual contact with animals, Beetz (2002) found that 1.8% of the men had had experiences with sadistic sex, 1.8% had actually used physical force in a sexual context without the agreement of the human partner, and 4.4% had practiced masochistic sex. With regard to sex with animals, 9.7% admitted to having used physical force.

Sandnabba et al. (2002) found that 7.4% of their participants, who practiced sadistic and masochistic sex with a consenting partner, had experienced sexual contact with animals. This sub-group also engaged in other unusual sex practices that were rare even among the sadomasochistic sample, and they were generally more sexually experimental, too. However, a general link between sadomasochistic practices and bestiality could not be affirmed; very few of the participants had engaged in both bestiality and sadomasochistic sex (Sandnabba et al. 2002). Unfortunately, this study did not address the question of whether sadistic or masochistic sexual practices were directly combined with bestiality.

In addition to the consenting sadomasochistic context, bestiality can also be found in cases of domestic violence where women are known to have been forced by their partners to perform sexual acts with animals (Walker 1979 cited in Ascione 1993; Dutton 1992). In another extreme form, the torturing or killing of an animal may serve as practice or surrogate for the killing of a human victim—a sexual connotation can be involved, but this does not necessarily have to be so. The "horse-ripper" cases are an example. Here, horses are injured, frequently in the genital area (Doninger 1993; A. Schedel-Stupperich, personal communication 2001), suggesting a sexual component of the abuse, even if the perpetrator did not perform explicit sexual acts.

The findings to date show that bestiality is more frequently practiced by persons who have either been sexually victimized or who themselves displayed sexual violence towards other persons. However, the prevalence rates are also quite remarkable in the normal population and in samples with different psychiatric problems. Thus, it cannot be argued that the practice of bestiality necessarily points to a history of sexual violence. Cases of bestiality need to be reviewed individually and may not be connected to any additional deviant behavior deserving intervention. Another complication in interpreting the findings on a link with interpersonal violence derives from the fact that few studies provide a detailed assessment of the behaviors subsumed under the term "bestiality" and do not distinguish between touching, oral contact, or penetration. Also, the degree of violence used is usually not assessed, probably because it is normally assumed that any kind of sexual contact with an animal represents, by definition, animal abuse. However, this information is of utmost importance if we are to fully understand this behavior. The motivation for sexual contact with an animal might be quite different between individuals. Information on any kind of force used will allow a differentiation of the quality of this act and certainly influence its public perception and predictive value. For example, "touching an animal's genitals" might be based on a curious exploration and be completely non-violent, while it could also be classified as one of many sexual behaviors carried out on an animal which incorporates some degree of force. It is very important to assess the degree of violence used with this behavior, especially when investigating violent offenders. Although bestiality is frequently defined as animal abuse, and much research indicates a link between interpersonal violence and animal abuse, it does not automatically follow that there is a link between non-violent forms of bestiality and interpersonal violence, as well. More detailed assessments might reveal a more complex situation. When approaching this problem from the other side and looking at persons who engage in sexual contact with animals and their anonymous reports of other violent behavior, the findings do not support a clear connection between sexual contacts with animals and violence, either. Taking into account factors such as the sexual practices engaged in with the animal, as well as the species involved—important characteristics of cases of bestiality—will allow a more informative differentiation and interpretation of such acts.

Types of Sexual Contact with Animals

Zoophilia and Bestiality

Commonly, the term "bestiality" is used to refer to a range of sexual contacts with animals, while sometimes it is more narrowly defined as penetrating, or being penetrated by, an animal (e.g., Home Office, Great Britain 2002). The term "zoophilia" is also widely used, predominantly by clinicians, but also by a group of people who engage in sex with animals and call themselves "zoophiles" or "zoos" (Beetz 2002; Miletski 2002). In contrast to bestiality, which just describes sexual contact, zoophilia is usually used to describe an exclusive or predominant desire for sexual contact with animals, as defined in DSM-IV (APA 1994), and also includes an emotional involvement with the animal. Zoophilia was first named in the DSM-III with the following diagnostic criteria: "The act or fantasy of engaging in sexual activity with animals is repeatedly preferred or the exclusive method of achieving sexual excitement." (APA 1980, p. 270). The revised edition of the DSM-III (DSM-III-R, APA 1987) added that zoophilia is nearly never a clinically significant problem by itself (Cerrone 1991). Like the DSM-III, the DSM-IV (APA 1994) subsumes zoophilia under the "paraphilias not otherwise specified (302.9)." It defines paraphilias in general as "recurrent, intense sexually arousing fantasies, sexual urges, or behaviors generally involving 1) nonhuman objects, 2) the suffering or humiliation of oneself or one's partner, or 3) children or other nonconsenting persons, that occur over a period of at least six months (Criterion A). … The behavior, sexual urges, or fantasies cause clinically significant distress or impairment in social, occupational, or other important areas of functioning (Criterion B)." (APA 1994, p. 523). The permanence and frequency of behaviors and fantasies can be employed as an indicator of the severity of the zoophilia (Arentewicz and Schmidt 1993). In contrast to this clinical definition, the persons who refer to themselves as "zoophiles" emphasize having an emotional involvement with the sexually desired animal, rather than the sexual activity itself (Beetz 2002; Miletski 2002). No information is available on the prevalence rates of zoophilia as defined by the DSM-IV or the self-designated group.

Sexual Practices

Sexual contact with animals can present itself in as many forms as it does with sexual interactions between humans. Masturbatory practices (rubbing against the body of the animal or masturbating male and female animals) have been reported, as well as oral contact, including mouth-to-mouth "kissing" and oral-genital contact carried out either by the human or the animal (Beetz 2002: Miletski 2002). Anal and vaginal penetration of the animal by men, anal penetration of men and women, and vaginal penetration of women by male animals occur, also (Beetz 2002: Miletski 2002: Williams and Weinberg 2003). And besides actively performing sexual acts with animals, watching others do it seems to be popular among a certain group of zoophiles.

Practices that can be considered violent or sadistic rather than mainly sexual include the insertion of objects into the anus or vagina of the animal, inflicting injuries to the animal or killing it. From case studies, there are reports of penetration of the cloaca of birds, mainly chicken or geese, which frequently led to severe injuries to the animals' internal organs and the death of the animal (Kosa et al. 1979). Sometimes, men cut the throat of birds, choke the animal, or break its neck while sexually penetrating them; the contractions caused by the dying bird increase the sexual excitement for the perpetrator (Bornemann 1990). Necrophilic tendencies have been revealed, too: animals are killed in connection with sexual gratification, and the dead body is used for masturbation, or is mutilated or dissected.

Some violent practices are performed with small rodents, such as mice or hamsters. Here, the animals are put into condoms or tubes and then inserted into the vagina or anus of the person. While the animal dies of suffocation, its movements stimulate the person (Miletski 2002). The insertion of fish or snakes into the vaginas of women has been reported by Dekkers (1994).

Many of the described sexual practices with animals can be seen in pornographic pictures and videos, and in some cases it is obvious that pain and suffering is inflicted upon the animal. Sedation of animals for the purpose of easier sexual abuse is sometimes practiced, too. Historically, live sex shows with animals were known to have existed in places such as Mexico, the Netherlands, and the US (Bagley 1968 cited in Miletski 2002), and supposedly still take place today (Beetz 2002). In these shows, male and female prostitutes engage in sexual activity, sometimes even intercourse, with mules, donkeys, or dogs (Ullerstam 1966; Love 1992).

Species Involved

A common assumption is that mainly farm animals are involved in cases of bestiality. This is probably based on early studies (e.g., Kinsey, Pomeroy and Martin 1948) which showed that bestiality occurred most often in rural areas. Also, pets are generally less accessible than livestock such as cows, calves, bulls, goats, sheep, and donkeys. Certainly, equines such as horses, donkeys, and mules, and ovines such as sheep or goats, seem to be frequently involved. Even sexual contact with bulls—masturbation of young bulls, and men and women being mounted and penetrated by bulls—has been reported (Masters 1966; Rosenfeld 1967). Chickens and rodents have also featured in the literature and are usually treated violently, leading to severe injuries and death of the animal. It has also been reported that humans engage sexually with male and female pigs, though not frequently, and supposedly even penetration of the human by male pigs occurs (Beetz 2002). Recent research (Beetz 2002; Miletski 2002; Williams and Weinberg 2003) shows that today, among samples of self-defined zoophiles, dogs, not just females of large breeds but also males, are preferred for sexual relations. And from an Internet survey, Rosenbauer (1997) found that the animals preferred for sexual relations were male dogs, followed by female horses, female dogs, and male horses.

Sexual contacts with deer, tapirs, antelopes, and camels have also been recorded (Massen 1994). And in former times, crocodile hunters in Egypt supposedly engaged in intercourse with female crocodiles before killing them (Massen 1994). Fish and snakes, too, are sexually used by inserting them into the vagina of the human (Dekkers 1994), and from Japan there are reports of penile penetration of dead thornback rays (Massen 1994). Cats can be employed for sexual stimulation by letting the animal lick the human genitalia, eat food from the vagina or penis (Miletski 2002), or by masturbating them. Chideckel (1935) and Rosenfeld (1967) provide accounts of women who practiced intercourse with monkeys, and Beetz (2002), Miletski (2002), and Williams and Weinberg (2003) found that some people engage sexually with dolphins.

With regard to sexual interactions with animals, the size and anatomy of the animal involved represent important factors to consider. Only animals of a certain size will not necessarily suffer pain or injuries from intercourse with a human male. Other practices, for example, masturbation or oral contact, can also be carried out without necessarily hurting the animal. However, in some cases the perpetrator does not care if his actions inflict injuries and pain to the animal, and in sadistic cases injuries, suffering or death of the animal are intentionally caused.

The Relevance of Bestiality to Veterinary Practice

Usually, veterinary practitioners will be confronted with the issue of sexual contact between humans and animals only if the animal is injured or an infection develops and the owner chooses to bring the animal in for an examination. Few studies have investigated the relevance of bestiality to the field of veterinary medicine, where, as in other professions, this subject is not easily addressed.

In 1972, Weidner questioned 400 veterinarians in Germany about cases of sexual and sadistic animal abuse seen in their practices. Thirty-six percent of the 294 veterinarians who replied had seen animals involved in bestiality in their own practice; the most frequently treated animals were cows and calves, but horses, pigs, sheep, fowl, and goats were also brought in. Relative to the population of each species in Germany at that time, horses were most often involved. Only a few cases of dogs were reported, and Weidner (1972) concluded that dogs, despite their growing popularity, were not targets of sexual approaches. Later research (Beetz 2002; Miletski 2002; Williams and Weinberg 2003) proved this to be a misconception. Cases of sexual contact with dogs, even penetration, that do not result in an injury are not likely to be noticed in a veterinary clinic, even if the dog is taken there for regular examinations (Beetz 2002). And if a sexual act leads to injuries, a veterinarian will often not be contacted because of the owner's fear and shame.

From her review of data on sadistic and sexual acts with animals in the literature, Weidner (1972) calculated that about 56% of the reported cases were sadistic and 29% were sexual; a combination of sadism and bestiality occurred in 15% of the cases. Only three men had caused injuries to the animal just through sexual acts; however, those had been severe. Weidner (1972) found that although intercourse could lead to irritation and minor injuries, perforations in the vaginal or rectal tissue were caused only by sadistic acts, and nearly always resulted in the death of the animal. Other negative reactions of the animal, such as diminished food-intake and nervousness, were also reported.

More recently, Munro and Thrusfield (2001, and in this issue of *Anthrozoös*) collected data from 404 UK veterinarians on animal abuse. Six percent reported cases classified as sexual abuse, with injuries in the genital and anal area, reproductive organs and surrounding tissue—in some cases caused by insertion of foreign bodies. In 21 cases dogs, and in three cases cats, had been injured. Overall, Munro and Thrusfield (2001) concluded that sexual animal abuse is a relevant problem encountered by veterinarians.

A discussion of bestiality as a problem for animal protection in Germany was provided by Stettner (1990). There the law on bestiality was abolished in 1969 and bestiality is not named explicitly in the paragraphs on animal protection. In law, sexual acts are not considered animal abuse per se, unless if serious or longer-lasting pain, suffering, or injuries to the animal are caused. According to Stettner (1990), acts like masturbation, oral-genital contact engaged in by the animal or by the human, or frottage are not an issue for animal protection, while intercourse may represent a relevant abuse, depending on the size of the animal and the degree of force used. Actions that definitely cause pain or injury to the animal, such as sadistic acts, insertion of objects which are too big, or similar, are prosecuted on the basis of the animal protection laws.

The Question of Consent

In cases of bestiality, it is usually assumed that violence or force has been exerted towards the animal and therefore the act constitutes animal abuse. However, it is increasingly acknowledged that sexual acts with animals can also be carried out without obvious force, and even with the animal as a seemingly willing participant (Beirne 2000; Beetz 2002). Nevertheless, there exists a widespread opinion that any form of bestiality is abusive because, similar to child abuse, it is impossible to obtain consent from animals (Ascione 1993; Beirne 2000). A very thorough and informative discussion of the question of consent has been provided by Beirne (2000). He proposes that any kind of sexual contact with an animal should be viewed as "interspecies sexual assault" rather than as bestiality, which has different definitions that include or exclude acts other than actual penetration of the animal.

Few authors report on the reactions of animals involved in sexual acts with humans but, as Beirne (2000) pointed out, these sometimes differ quite remarkably. At one extreme, some animals, such as dogs, seem to enjoy the attention provided by the sexual interaction with a human (Greenwood 1963 cited in Miletski 2002; Blake 1971; Beirne 2000) or initiate the sexual interaction themselves (Dekkers 1994). At the other extreme, fish, fowl, or other small animals obviously suffer and are unwilling recipients of sexual advances by humans (Beirne 2000). Between these extremes lie more neutral reactions or consequences for the animal, as seen in pornographic videos (Beirne 2000). Here, the reactions of some cows, horses, or other large quadrupeds seemed close to boredom and indifference, or perhaps "calculated detachment," as animals such as these are also put through many other invasive procedures by humans, for example, medical procedures and pregnancy testing.

According to Beirne (2000), however, all sexual acts between humans and animals, independent of the use of force or the reaction of the animal, are wrong. One argument is that in almost every situation the relationship between humans and animals carries the potential of coercion (Adams 1995 cited in Beirne 2000), simply because domestic pets as well as livestock depend on humans for food, shelter, and affection, and humans have the capacity to catch wild animals and to subject them to their will. This potential for coercion applies also to sexual activities. Drawing a parallel to sexual assault of women or children, Adams (1995 cited in Beirne 2000) stated that bestiality almost always involves forced sex because it is obtained by physical, economic, psychological, or emotional coercion, and in such cases it is impossible to obtain consent. However, not the imbalance of power, but the fact that one of the involved parties does not, or cannot, give or communicate genuine consent to sexual relations is the decisive criterion (Beirne 2000).

It might seem that a lack of resistance could be taken as consent, but that does not mean genuine consent is present. To be able to give genuine consent, both parties involved must be conscious, fully informed, and positive in their desires (Beirne 2000). Beirne (2000) demands that if these criteria apply to human–human sexual relations they should also apply to humans' sexual advances to animals, given that animals are sentient beings. However, applying human criteria, such as the demand for being fully informed, to animals seems rather anthropocentric and not a useful attempt to understand and apply criteria more suitable for the perceptive and cognitive abilities of animals. According to Beirne (2000), bestiality must be condemned because animals cannot give consent in a way humans can readily understand, or cannot resist humans in any significant way due to their sometimes docile nature or other human-bred features. Even if the animal communicates protest by scratching, biting, or making sounds, usually the animal will be unable to effectively resist (Beirne 2000). However, the poor ability of some animals to effectively resist should not be the issue: rather, the issue should be the unwillingness or inability of the human to perceive, understand and react accordingly to resistance.

The question of where the line between non-sexual and sexual actions is to be drawn was left unanswered by Beirne (2000). While he viewed the milking of cows as clearly non-sexual, with other actions he questioned if the intent should be taken into consideration, for example, with electrically or manually induced ejaculation for insemination, an objectively sexual act, or artificial insemination (Beetz 2004).

Besides his discussion of consent, Beirne (2000) proposed a typology of interspecies sexual assault, distinguishing between sexual fixation, commodification, adolescent sexual experimentation, and aggravated cruelty. With regard to a sexual fixation like zoophilia, Adams (1995) stated that many zoophiles—just like child abusers and rapists—claim that the sex is consensual and benefits the sexual partners as well as themselves (Adams 1995); Beirne (2000) was not aware of any data that would confirm this view. However, recent research provides some support for Adams' claim (Beetz 2002; Miletski 2002; Williams and Weinberg 2003), although these studies do not favor a comparison of zoophilia with child molestation and rape. Rather, the studies showed that there are people who engage in sex with animals who emphasize the emotional side of the relationship, and who claim not to use any form of coercion.

A new and important feature of Beirne's (2000) definition of bestiality as interspecies sexual assault is that he moved away from a mainly anthropocentric view that condemns bestiality because of societal, religious, or moral concerns, to an approach that focuses on the rights and welfare of the animal. According to Beirne (2000), human–animal sexual relations are wrong because 1) they nearly always involve coercion, 2) they often cause pain and death, and 3) animals are unable to communicate consent in a form that humans can readily understand or speak out about their abuse. Beirne (2000) stated that he did not want to promote the criminalization or psychiatrization of bestiality, but thought that if bestiality was regarded as an objectionable act, then the human perpetrator should be "censured." However, he did not seem to favor penalties as harsh as those found in former times.

Depending on the country, bestiality was or still is viewed with "moral, judicial and aesthetic outrage" (Beirne 2000, p. 314) and many people react to its mentioning with bewilderment or disgust, rather than intellectual inquiry. Nonetheless, in the nineteenth century many "unnatural offences" like homosexual contact or bestiality were decriminalized and a more liberal stance towards unusual practices has spread. In the US, no federal law on bestiality exists and today only about half of its states have bestiality laws (Beirne 2000). Germany abolished its law on bestiality in 1969, and other European states also do not criminalize every sexual contact with animals. Societal control of bestiality has passed from religion and criminal law to the medical profession of psychiatry, and there is even a degree of dismissal, if not tolerance. Beirne (2001) called this a pseudoliberal tolerance of bestiality and criticized the philosopher Peter Singer (2001), well-known for his work Animal Liberation (1975), for his review of Midas Dekker's book Dearest Pet, one of the few informative books on human–animal sexual relations. Singer (2001) stated that the cause of the existing taboo against bestiality, and the vehemence with which it is held, is the desire of humans to differentiate themselves from animals, erotically and in other ways. For Singer, sexual relations with animals are to be condemned if they involve cruelty.

The issue of consent and the criminal status of bestiality has become a topic that is now repeatedly discussed among animal protectionists, as well as zoophiles (Beetz 2002, 2004). The Humane Society of the United States (HSUS 1999) started a campaign against sexual abuse of animals which, according to the HSUS, is the eroticization of violence, control, and exploitation. The HSUS regards all sexual contacts with animals as abuse, following the arguments given by Ascione (1993) and Beirne (2000) that animals lack the ability to give and communicate consent. In contrast to this, zoophiles themselves question this inability of animals to refuse sexual advances in a way that humans can understand, if the humans are willing and sensitive enough to perceive and interpret the signs correctly, and to act accordingly (Beetz 2002). Obviously, this is not consent as defined by law, neither is it conscious consent by a sentient being that is fully informed and positive in its desires. It is probably more difficult to establish the consciousness and the state of being fully informed with animals than it is to deduce from behavior whether an animal is positive in its desires or is at least not resisting. But it is important to note that many zoophiles do not regard non-resistance as sufficient evidence for consent (Beetz 2002). Instead, they also place great importance on positive reactions from the animal, such as approaching the person, cuddling, rubbing against the person, not trying to move away, and displaying sexual excitement (Beetz 2002). Some zoophiles claim that they carefully react to the nonverbal signs of the animals and some even try to communicate with the animal, as described by Masters (1966), for example.

However, to provide a perspective to this discussion of consent in relation to bestiality, other situations need to be considered where an animal's consent is obviously regarded as less important. Everybody agrees that if an animal is deliberately hurt or killed by someone for personal pleasure, then this represents animal abuse. But opinions are divided if the killing or hurting happens in the service of society, for example, slaughtering animals for food or using them in medical research. In these cases, few people stress the question of consent. For the animal, the motivation of the person behind the act very likely makes no difference—it would not consent to any act that causes pain or death.

Humans still deal with many animal-related questions with a very anthropocentric view, and as soon as sexuality is involved, reactions become less rational and more emotional and moral. The involvement of sexuality and gain of sexual pleasure for the human makes an important difference in the public opinion on bestiality: it goes against people's sense of decency. But animals do not hold the human cultural or societal values that say an act is disagreeable just because it is sexual. From the perspective of animals, their own sexual excitement and absence of psychological and physical stress, pain, injury, or death are probably more important.

Interestingly, procedures that definitely have sexual characteristics, such as the electrical collection of semen, pregnancy testing or artificial insemination, are widely accepted. And yet, these interactions are probably more uncomfortable than the non-forceful insertion of a human penis or manual masturbation with the intention of personal sexual pleasure (Beetz 2004). Also, a procedure such as neutering has a strong impact on an animal's sexuality, and yet is widely accepted and carried out frequently without raising the question of consent.

It is also questionable whether humans really cannot understand at least a little of what an animal tries to communicate, even if this requires some knowledge of the behavior of different species, and perhaps an ability for deduction. It has never been doubted by most animal owners that they know at least some of the likes and dislikes of their own pets, and that they can deduce this from their animal's behavior. Nevertheless, I agree that this approximates, rather than equals, a clear communication of consent.

It has to be added to this discussion of consent (in a legal or non-legal sense) that certainly many people engaging in sexual contact with animals either do not have the ability to understand the nonverbal communication of the animal or just do not care about the animal's welfare. They may even deliberately inflict pain and injuries. Opinions about consent in regard to bestiality, however, seem to be influenced much more by emotion and moral attitudes than opinions about other non-sexual issues involving animals. Consent is also not considered important for most of the legislation on bestiality. Still, the main reasons for its criminalization seem to be concerned with morals, together with the possible danger to society due to an association of bestiality with other sexually deviant behavior, in particular sex offending.

Discussion

The information presented in this paper shows that understanding sexual contact with animals, and its relationships with violence towards animals and people and sex offending, is not easy. A first step is to acknowledge that bestiality does not occur as infrequently as many people would like to believe. It always has been, and still is, a behavior engaged in by a significant number of people. The data also show that bestiality is found most often among sexually abused persons, among violent offenders (especially those who are highly aggressive), and among sex offenders (especially those who employ more violence in their offenses). This suggests the existence of a connection between bestiality and violence, but already it becomes obvious that the degree of violence used in the offense seems to play a role. Unfortunately, this is seldom assessed in detail. If such information was available, it seems likely we would find an even stronger association between more violent acts towards humans and more violent acts towards animals, sexual or non-sexual. These data need to be obtained to see if there really is a link between interpersonal violence and all forms of sexual contact with animals, or if it is only associated with violent cases of bestiality.

As has been mentioned before, sexual acts often include practices that do not necessarily employ coercion, restriction, force, or violence; not only oral-genital contact or masturbation, but also penetration of the animal can be practiced without causing pain or injuries. Especially with the animal as an active partner, including cases where women sexually approach male animals, techniques that sexually arouse the animal (rather than scare) seem necessary for the intended purpose. The acceptance that these acts can be non-violent leads to a questioning of the simple association between sexual contact with animals in general and interpersonal violence.

Sexual contact with animals has only recently been investigated in more detail by Beetz (2002), Miletski (2002) and Williams and Weinberg (2003). Their data show that there is a group of people who engage in sexual interactions with animals without employing force or violence. Rather, they try to sexually arouse the animal to reach their goal. It is clear that not all persons practicing bestiality are like this, and it needs to be acknowledged that there is a wide range of practices, intentions, and degrees of force used in these contacts. However, even in regard to non-violent encounters, authors such as Lockwood and Ascione (1998) and Beirne (2000) point to the fact that animals cannot give genuine consent, in a legal sense. This is a valid argument, but this factor is probably more important in the discussion of the criminalization of sexual contacts with animals than in the attempt to find a link with interpersonal violence and sex offending. The parallels that are often drawn with sexual abuse of children relate to this question of consent, but no research exists that really proves that the same motivations apply to child sexual abuse and bestiality. Moreover, in cases of bestiality, the animals are usually sexually mature and show a certain sex drive that would be directed to an animal partner, if available. So even if consent is not obtained, the quality of the act—if the animals suffers, enjoys it, or reacts with indifference—provides important information that needs to be included in any evaluation of a case.

With regard to legislation, bestiality is a difficult subject. Depending on the country, sexual contact with animals can be legal or illegal, and punished either mildly or with imprisonment. Most of the current laws are part of legislation on sex offending, as in the US and the UK, implying that bestiality is not only a sexual offense but one that also needs to be addressed because of its potential danger to society and violation of religious and moral codes. A less anthropocentric view, focusing more on the welfare and protection of the animal, would be useful. Of course, defining bestiality as animal abuse probably does not promote a critical and detailed assessment of individual cases in court. Overall, a more promising approach would be to handle cases of bestiality according to animal protection laws, assessing the degree of violence and force and underlying motivations, and using psychiatric evaluations of violent cases to discover whether there are grounds for applying sex offender laws. With this approach, relatively harmless acts such as masturbating a male animal, initiating oral-genital contact or similar, for example, in cases of juvenile experimentation, could be addressed without criminalizing, humiliating and inflicting severe societal consequences on the person involved. Animal protection law needs to be applied in sexual animal abuse cases without focusing too much on the fact that sexuality is involved, except in sadistic cases. As with other forms of abuse, there is need for research into intentional sexual abuse involving violence and force, non-caring attitudes towards living beings, and especially into cases that reveal pleasure was gained from inflicting pain on animals.

Acknowledgements

I thank the reviewers for their helpful advice.

References

Adams, C. J. 1995. Woman-battering and harm to animals. In *Animals and Women: Feminist Theoretical Explorations*, 55–84, ed. C. Adams and J. Donovan. Durham, NC: Duke University Press.

Alvarez, W. A. and Freinhar, J. P. 1991. A prevalence study of bestiality (zoophilia) in psychiatric in-patients, medical in-patients, and psychiatric staff. *International Journal of Psychosomatics* 38: 45–47.

APA. 1980. *Diagnostic and Statistical Manual of Mental Disorders.* 3rd edn. Washington, DC: American Psychiatric Association.

APA. 1987. *Diagnostic and Statistical Manual of Mental Disorders* 3rd edn. revised. Washington, DC: American Psychiatric Association.

APA. 1994. *Diagnostic and Statistical Manual of Mental Disorders* 4th edn. Washington, DC: American Psychiatric Association.

Arentewicz, G. and Schmidt, G. eds. 1993. *Sexuell gestörte Beziehungen.* Stuttgart: Ferdinand Enke.

Arluke, A. 1998. The relationship between animal cruelty and subsequent anti-social behavior. Paper presented at the 8th International Conference on Human–Animal Interactions, Prague, 10–12th September.

Arluke, A., Levin, J., Luke, C. and Ascione, F. 1999. The relationship of animal abuse to violence and other forms of antisocial behavior. *Journal of Interpersonal Violence* 14(9): 963–975.

Ascione, F. R. 1993. Children who are cruel to animals: A review of research and implications for developmental psychopathology. *Anthrozoös* 6(4): 226–247.

Ascione, F. R. 1999. The abuse of animals and human interpersonal violence. In *Child Abuse, Domestic Violence, and Animal Abuse*, 50–61, ed. F. R. Ascione and P. Arkow. West Lafayette, IN: Purdue University Press.

Ascione, F. R. and Arkow, P. eds. 1999. *Child Abuse, Domestic Violence, and Animal Abuse. Linking the Circles of Compassion for Prevention and Intervention.* West Lafayette, IN: Purdue University Press.

Ascione, F. R., Friedrich, W. N., Heath, J. and Hayashi, K. 2003. Cruelty to animals in normative, sexually abused, and outpatient psychiatric samples of 6- to 12-year-old children: Relations to maltreatment and exposure to domestic violence. *Anthrozoös* 16(3): 194–212.

Bagley, H. L. 1968. *The Beast Seekers.* Atlanta, GA: Pendulum Books.

Beetz, A. 2002. *Love, Violence, and Sexuality in Relationships between Humans and Animals*. Doctoral dissertation. Aachen, Germany: Shaker Verlag.

Beetz, A. 2004. Bestiality/ zoophilia: A scarcely investigated phenomenon between crime, paraphilia and love. *Journal of Forensic Psychology Practice* 4(2): 1–36.

Beirne, P. 2000. Rethinking bestiality: Towards a concept of interspecies sexual assault. In *Companion Animals and Us: Exploring the Relationships between People and Pets*, 313–331, ed. A. L. Podberscek, E. S. Paul and J. A. Serpell. Cambridge: Cambridge University Press.

Beirne, P. 2001. Peter Singer's "Heavy Petting" and the politics of animal sexual assault. *Critical Criminology* 10: 43–55.

Blake, R. 1971. *Beauty/Beast: A Study of Animal Eroticism, Vol. 2.* 2nd edn. San Diego, CA: Greenleaf Classics, Inc.

Bornemann, E. 1990. *Enzyklopädie der Sexualität.* Frankfurt/Berlin: Ullstein.

Cerrone, G. H. 1991. Zoophilia in a rural population: two case studies. *Journal of Rural Community Psychology* 12(1): 29–39.

Chideckel, M. 1935. *Female Sex Perversions.* New York: Eugenics Publishing Company.

Dekkers, M. 1994. *Dearest Pet: On Bestiality.* New York: Verso.

DeViney, E., Dickert, J. and Lockwood, R. 1983. The care of pets within child abusing families. *International Journal for the Study of Animal Problems* 4: 321–329.

Doninger, W. 1993. Diary. *London Review of Books*. 23 September: 25.

Dutton, M. A. 1992. *Empowering and Healing the Battered Woman.* New York: Springer.

English, K., Jones, L., Patrick, D. and Pasini-Hill, D. 2003. Sexual offender containment—use of the Postconfiction Polygraph. *Annals of the. New York Academy of Science* 989: 411–427.

Felthous, A. R. and Kellert, S. R. 1986. Violence against animals and people: is aggression against living creatures generalized? *Journal of the American Academy of Psychiatry and the Law* 14: 55–69.

Felthous, A. R. and Kellert, S. R. 1987. Childhood cruelty to animals and later aggression against people: A review. *Journal of Psychiatry* 144: 710–717.

Fleming, W. M., Jory, B. and Burton, D. L. 2002. Characteristics of juvenile offenders admitting to sexual activity with nonhuman animals. *Society & Animals* 10: 31–45.

Flynn, C. P. 1999. Exploring the link between corporal punishment and children's cruelty to animals. *Journal of Marriage and the Family* 61: 162–177.

Frazier, M. R. 1997. Physically and sexually violent juvenile offenders: A comparative study of victimization history variables. Ph. D. dissertation, Utah State University, Logan.

Friedrich, W. N., Grambsch, P., Damon, L., Hewitt, S. K., Koverola, C., Lang, R. A., Wolfe, V. and Broughton, D. 1992. Child Sexual Behavior Inventory: Normative and clinical comparisons. *Psychological Assessment* 4: 303–311.

Gebhard, P. H., Gagnon, J. H., Pomeroy, W. B. and Christenson, C. V. 1965. *Sex Offenders: An Analysis of Types.* New York: Harper and Row, and Paul B. Hoeber, Inc. Medical Books.

Gleyzer, R., Felthous, A. R. and Holzer III, C. E. 2002. Animal cruelty and psychiatric disorders. *Journal of the American Academy of Psychiatry and the Law* 30: 257–265.

Greenwood, M. D. 1963. *Unusual Sex Practices.* Hollywood, CA: The Genell Corp.

Hellman, D. S. and Blackman, N. 1966. Enuresis, firesetting and cruelty to animals: a triad predictive of adult crime. *American Journal of Psychiatry* 122: 1431–1435.

Holmes, R. M. 1991. *Sex Crimes*. Newbury Park: Sage Publications.

Home Office, Great Britain 2002. *Protecting the Public. Strengthening Protection against Sex Offenders and Reforming the Law of Sexual Offences.* Norwich: HMSO.

Humane Society of the United States (HSUS) 1999. Animal Sexual Abuse Fact Sheet. http:// www. hsus.org/firststrike/factsheets/facts.html.

Hunt, M. 1974. *Sexual Behavior in the 1970s.* Chicago, IL: Playboy Press.

Kinsey, A. C., Pomeroy, W. B. and Martin, C. E. 1948. *Sexual Behavior in the Human Male*. Philadelphia: Saunders.

Kinsey, A. C., Pomeroy, W. B., Martin, C. E. and Gebhart, P. H. 1953. *Sexual Behavior in the Human Female.* Philadelphia: Saunders.

Kosa, R., Vargha, M., Viragos-Kis, E. and Szilard, J. 1979. Sodomie (Zoophilie, Bestialität). Ein Beitrag zu ihren Konsequenzen für die Persönlichkeitsentwicklung und den Gefahren dieser sexuellen Aberration für die Gesellschaft an Hand von Fällen aus der Praxis. *Archiv Kriminologie* 164(1-2): 36–44.

Lewis, D. O., Shanok, S. S., Grant, M. and Ritvo, E. 1983. Homicidally aggressive young children : neuropsychiatric and experiential correlates. *American Journal of Psychiatry* 140: 148-153.

Lockwood R. and Ascione F. R. eds. 1998. *Cruelty to Animals and Interpersonal Violence: Readings in Research and Application.* West Lafayette, IN: Purdue University Press.

Love, B. 1992. *Encyclopedia of Unusual Sex Practices.* Fort Lee, NJ: Barricade Books.

Massen, J. 1994. *Zoophilie. Die sexuelle Liebe zu Tieren.* Köln: Pinto Press.

Masters, R. E. L. 1966. *Sex-Driven People*. Los Angeles, CA: Sherbourne Press, Inc.

McClellan, J., Adams, J., Douglas, D. McCurry, C. and Storck, M. 1995. Clinical characteristics related to severity of sexual abuse: A study of seriously mentally ill youth. *Child Abuse and Neglect* 19: 1245–1254.

Merz-Perez, L. Heide, K. M. and Silverman, I. J. 2001. Childhood cruelty to animals and subsequent violence against humans. *International Journal of Offender Therapy and Comparative Criminology* 45: 556–573.

Miletski, H. 2002. *Understanding Bestiality and Zoophilia.* Bethesda, MD: Author.

Miller, K. S. and Knutson, J. F. 1997. Reports of severe physical punishment and exposure to animal cruelty by inmates convicted of felonies and by university students. *Child Abuse and Neglect* 21(1): 59–82.

Munro, H. M. C. and Thrusfield, M. V. 2001. "Battered pets": sexual abuse. *Journal of Small Animal Practice* 42: 333–337.

Ressler, R. K., Burgess, A. W. and Douglas, J. E. 1988. *Sexual Homicide: Patterns and Motives.* Lexington, MA: Lexington Books.

Ressler, R. K., Burgess, A. W., Hartmann, C. R., Douglas J. E. and McCormack, A. 1986. Murderers who rape and mutilate. *Journal of Interpersonal Violence* 1: 273–287.

Rosenbauer, F. 1997. *Sexueller Kontakt mit Tieren.* Semesterarbeit in der Veranstaltung "Sexuelle Störungen" im SS 1997 bei Prof. C. Möller. Siegen.

Rosenfeld, J. R. 1967. *The Animal Lovers.* Atlanta, GA: Pendulum.

Sandnabba, N. K., Santtila, P., Beetz, A. M., Nordling, N. and Alison, L. 2002. Characteristics of a sample of sadomasochistically-oriented males with recent experience of sexual contact with animals. *Deviant Behavior* 23(6): 511–530.

Schiff, K., Louw, D. and Ascione, F. R. 1999. Animal relations in childhood and later violent behaviour against humans. *Acta Criminologica* 12: 77–86.

Schweitzer, I. and Beetz, A. M. 2000. Special evaluation of data on sexual fantasies collected by Schweitzer. Diplomarbeit, Friedrich-Alexander-Universität. Erlangen-Nürnberg, Germany.

Singer, P. 1975. *Animal Liberation.* New York: Avon.

Singer, P. 2001. Heavy petting. *Nerve* (March/April). www.nerve.com/Opinions/Singer/heavyPetting.

Stettner, M. 1990. Unzucht mit Tieren—ein Tierschutzproblem. *Deutsche Tierärztliche Wochenschrift* 97: 137–192.

ten Bensel, R. W., Ward, D. A., Kruttschnitt, C., Quigley, J. and Anderson, R. K. 1984. Attitudes of violent criminals towards animals. In *The Pet Connection: Its Influence on Our Health and Quality of Life*, 309–318, ed. R. K. Anderson, B. L. Hart and L. A. Hart. Minneapolis, MN: CENSHARE, University of Minnesota.

Tingle, D., Barnard, G. W., Robbins, L., Newman, G. and Hutchinson, D. 1986. Childhood and adolescent characteristics of pedophiles and rapists. *International Journal of Law and Psychiatry* 9: 103–116.

Ullerstam, L. 1966. *The Erotic Minorities.* New York: Grove Press.

Vermeulen H. and Odendaal, S. J. 1993. Proposed typology of companion animal abuse. *Anthrozoös* 6: 248–257.

Walker, J. R. 1979. *The Battered Woman.* New York: Harper and Row.

Weidner, E. 1972. Sodomie und Sadismus als Tierschutzproblem. Doctoral Dissertation. University of Giessen, Germany.

Williams J. W. and Weinberg M. S. 2003. Zoophilia in men: A study of sexual interest in animals. *Archives of Sexual Behavior* 32(6): 523–535.

"Battered pets:" sexual abuse++

H. M. C. Munro† and M. V. Thrusfield**

* Department of Veterinary Clinical Studies, University of Edinburgh, Scotland, UK

† Department of Veterinary Pathology, University of Edinburgh, Scotland, UK

Abstract

A study of non-accidental injury in small animals in the UK, based on responses from a random sample of small-animal practitioners, identified six percent of the 448 reported cases as being sexual in nature. Twenty-one cases occurred in dogs, five in cats and two in unspecified species. Reasons for suspecting sexual abuse were: the type of injury; behavior of the owner; statements from witnesses; and admission by the perpetrator. Types of injury included vaginal and anorectal penetrative (penile and non-penile) injury, perianal damage, and trauma to the genitals. Some injuries (such as castration) were extreme, and some were fatal. In contrast, other cases revealed no obvious damage. The type and severity of injuries were similar to those described in texts on child abuse and human forensic pathology.

Keywords: *companion animals, injuries, sexual abuse*

This is the third in a series of papers describing a study of physical abuse in dogs and cats—that is, abuse by deliberate physical injury (so-called non-accidental injury [NAI] or 'battered pets'—by analogy with the well documented "battered-child syndrome" [Kempe et al. 1962]). The study was based on the experiences of a sample of small animal practitioners in the UK. The first paper published in this series addressed the features that raised suspicion, or allowed recognition, of such abuse (Munro and Thrusfield 2001a), while the second detailed the injuries involved (Munro and Thrusfield 2001b).

The study focussed primarily on physical abuse (NAI), but also identified cases of a sexual nature. Vermeulen and Odendaal (1993) included one aspect of sexual abuse, "bestiality," as a form of physical abuse (NAI) in their proposed typology of companion animal abuse (without documenting details of cases), but, conventionally, cases of sexual abuse are classified separately from NAI (Meadow 1997; Munro and Thrusfield

Address for correspondence: Helen Munro, Royal (Dick) School of Veterinary Studies, University of Edinburgh, Summerhall, Edinburgh, Scotland, UK. EH9 1QH E-mail: Helen.Munro@ed.ac.uk.

++ This paper was originally published in *Journal of Small Animal Practice* (2001), 42: 333–337. Reprinted with permission.

2001a). However, it was considered desirable to present a detailed account of sexual abuse cases in this series, albeit under the general series heading "battered pets," because, to the best of the authors' knowledge, it is the first published report of such cases, and therefore complements the study in what is probably its most sensitive area.

Terminology

Further clarification of the term "sexual abuse," and other associated terms, with regard to sexual activity between humans and animals, is essential. The definition of the commonly used term "bestiality" has varied over the centuries, and in different parts of the world (Beirne 1997). Its modern definition, and that of its less well-known, etymologically anodyne synonym, "zoophilia" (Greek:ζωο = animal, -φιλία = affection), is "sexual activity between a person and an animal" (Hanks 1986). However, if "bestiality" is mentioned in the literature, it is usually confined to one or two lines. Thus, Gee and Watson (1991), in their text on sexual offences, summarily dismiss it as "sexual intercourse with animals, a form of sexual perversion obviously more likely to be encountered in the country" (implying that it is generally confined to livestock), and Tannahill (1989), in her extensive study of sex in history, gives it no more than a cursory mention. Moreover, neither expression reflects the wide spectrum of sexually motivated acts that may be involved, nor do they draw attention to any actual physical injury that may be caused—a fact appreciated by Beirne (1997) who points out that the range of animals used in sexual activity is wide, as is the spectrum of injury.

In humans, sexual abuse (using this term to include not only the sexual abuse of children but also sexual offences against adults of both sexes) may involve the anorectal region as well as the genitalia of both sexes. The injuries may vary from none (as is commonly noted in child sexual abuse [Hobbs, Hanks and Wynne 1999a]) to extreme, where violent acts cause death (Gee 1985). Such detail has been lacking in the animal abuse field until now, and in this study, all the cases, with only two exceptions, were reported to involve physical injury, some of which was extreme.

With regard to the definition of activities that appear similar to those inflicted on people, but involve the use of animals, the term "animal sexual abuse" (following the contemporary practice of using nouns as adjectives, which is also consistent with the accepted term "child sexual abuse") is therefore more precise and encompassing than either bestiality or zoophilia. Beirne (1997) offers an alternative term, "interspecies sexual assault," but this is not favored by the authors because it could be interpreted to imply improbable activities (the sexual abuse of one animal species by another).

Materials and Methods

Sampling of Veterinarians

An anonymous questionnaire, eliciting details of veterinarians' perceptions and experience of NAI, was distributed to a sample of 1,000 veterinarians, randomly selected from a list of members of the British Small Animal Veterinary Association (BSAVA) resident in the UK. The four sections of the questionnaire recorded information on: (1) acknowledgement and recognition of NAI; (2) individual cases (species, breed, age, gender, reasons for suspicion/recognition, nature of injuries, and outcome for the animal); (3) experience of putative characteristic features (for example, history inconsistent with injury); and (4) any salient comments. Data were stored on a custom-built database, using Access 97 (©Microsoft, Redmond, WA). Full details of the sampling procedure are presented in the first paper in this series (Munro and Thrusfield 2001a).

Definition of Cases of Sexual Abuse

Cases were defined as involving sexual abuse if they:

- had been explicitly identified as such by respondents;

or:

- displayed characteristics that would classify them as sexual abuse cases in the human literature (as outlined in the introduction), but without explicit identification by respondents.

Results

Questionnaires were returned by 404 respondents, and detailed abuse in 243 dogs, 182 cats, three horses, two rabbits, one hamster, two "dogs and cats" and 15 unspecified species. These included 21 cases of sexual abuse in dogs (14 females and seven males), and five cases in cats (three females, one male and one of unspecified gender) (Table 1). In addition, two respondents reported injuries of a sexual nature (cases 27 and 28), without recording either species or gender (although one—case 27—was obviously female, because the injuries were unexplained vaginal injuries). Two respondents reported two cases each. All others reported a single case.

Table 1 documents fully the case reports submitted by the respondents. Cases are grouped (as far as possible) according to the main sites and type of injury. For example, cases 1 to 5 list known, or suspected, penile penetrative abuse, involving the vagina in dogs, and cases 6 to 13 report various general vulval, vaginal and uterine injuries in dogs.

In six cases (cases 1 to 5 and case 16), the nature of the injuries, and related circumstances, allowed respondents to identify "sexual abuse"

Table 1. Individual cases of known and suspected sexual abuse in dogs and cats

Case	Species	Breed	Age	Gender	Details Reported by Respondent	Outcome
1	Dog	Border collie	> 2 yrs.	F	Vaginal injuries present. Wife saw husband having sex with the dog	Survived
2	Dog	Crossbreed	> 2 yrs.	FN	Owner claimed dog sexually abused by lodger, who pled guilty and asked for other offences to be taken into account. Respondent reported no injuries noted	Survived
3	Dog	Dalmatian	NR	F	Sudden onset bleeding from vulva. Trauma to vagina. Dog possibly "raped" by human	Survived
4	Dog	Cocker spaniel	NR	F	Recurrent refractory vaginitis. In-depth investigation undertaken, which suggested the cause was "human interference"	Survived
5	Dog	Dalmatian	NR	F	Owner requested examination of bitch because she suspected that her husband had "interfered" with the dog. The bitch was in season, showing pro-estral bleeding, and was externally normal. No internal examination was made and no samples taken	Survived
6	Dog	Crossbreed	NR	F	Gross vaginal injuries caused suspicion	Survived
7	Dog	Crossbreed	> 2 yrs.	F	Hemorrhage from vulva. Post-mortem examination revealed knife wound deep in vagina. Blood vessels in abdomen had been severed	Died
8	Dog	Crossbreed	> 2 yrs.	F	Sharp point felt on abdominal palpation was the sharp end of a knitting needle, whose other end was vaginal in location. The 12 inch long needle had penetrated uterine/cervical wall	Survived
9	Dog	Staffordshire bull terrier	NR	F	Tear in uterus just rostral to the cervix, with no "natural" explanation for injury. Track of scarring through cervix, with puncture just anterior, through uterus. Young girl in household suspected of causing injury	Survived
10	Dog	Crossbreed	7 mo. to 2 yrs.	F	Intra-cervical/vaginal foreign body (candle) found during routine ovariohysterectomy of stray dog	Survived
11	Dog	Staffordshire bull terrier	> 2 yrs.	F	Multiple hemorrhages around vulva, and internally (around uterus and within abdominal wall)	Survived

NR Not recorded, M Male, F Female, N Neutered

Individual cases of known and suspected sexual abuse in dogs and cats…continued

Case	Species	Breed	Age	Gender	Details Reported by Respondent	Outcome
12	Dog	Collie	> 2 yrs.	F	Piece of broom handle recovered from distal vagina. Secondary infection of vaginal canal	Survived
13	Dog	English bull terrier	7 mos. to 2 yrs.	F	Laparotomy for abdominal pain/suspected peritonitis. Rostral vagina punctured by piece of stick, a fragment of which was still present at the puncture site. Investigation was undertaken by an animal welfare charity, and male family member then admitted abuse	Survived
14	Dog	Lurcher	7 mos. to 2 yrs.	F	Broomstick inserted into rectum up to level of liver	Died
15	Dog	Yorkshire terrier	> 2 yrs.	M	Mucosa all around anus damaged and almost necrotic. No obvious explanation for lesion	Survived
16	Dog	Crossbreed	> 2 yrs.	M	Dog had been seen being sexually abused. Anal ring dilated and both ears chronically thickened. Respondent had not at first believed the abuse, but several people were aware that it was occurring and had reported a man to the police	Survived
17	Dog	Crossbreed	> 2 yrs.	M	Cord ligature tied round base of penis.	Survived
18	Dog	Collie	> 2 yrs.	M	Elastic band had been placed around base of dog's scrotum, which was swollen and necrotic. Single episode but more suspected when owner interviewed. Perpetrated by teenage sons	Survived
19	Dog	Labrador	7 mos. to 2 yrs.	M	Severe, necrotizing swelling of scrotum and testes, with evidence of constriction high on scrotum, consistent with a broad ligature which had been removed. Foster child in household later admitted having applied a rubber band to scrotum	Survived
20	Dog	Red setter	7 mos. to 2 yrs.	M	The dog was found alone, hemorrhaging profusely, and had been castrated with a very sharp implement, leaving an open wound. Dog was not reclaimed	Survived
21	Dog	Collie cross	> 2 yrs.	M	Large, penetrating pararectal wound. Respondent reported that appearance of injury, and owner, had caused suspicion	Survived

NR Not recorded, M Male, F Female, N Neutered

Individual cases of known and suspected sexual abuse in dogs and cats…continued

Case	Species	Breed	Age	Gender	Details Reported by Respondent	Outcome
22	Cat	NR	NR	NR	Deep incised wound in rectum and perineum. Respondent noted that it was difficult to see how this could have been accidental	Survived
23	Cat	Domestic shorthaired	7 mos to 2 yrs	FN	Deep single penetrating wound, thought to be stab wound, caused by a "large sharp instrument," in perivulval and perianal area. Respondent considered the wound "far too clean and neat to be an animal tooth/claw wound"	Survived
24	Cat	Domestic shorthaired	7 mos.. to 2 yrs	F	Foreign object, bloody and infected, had been inserted into the vagina. Thought to be tampon. Considered unlikely to have occurred accidentally	Survived
25	Cat	Domestic shorthaired	7 mos. to 2 yrs	FN	Four equidistant, radial splits*—consistent with the insertion of large object—present in anus. Respondent considered lesion was inexplicable, and noted owner angry on questioning	Survived
26	Cat	Domestic	< 12 wks.	M	Traumatized genitalia. Kitten soaking wet, with a strong, sweet smell, which resembled after-shave lotion or shampoo. Teenage perpetrator also violent to his mother‡	Died
27	NR	NR	NR	NR	Unexplained vaginal injuries	NR
28	NR	NR	NR	NR	Perianal wounds, consistent with stabbing. Described by respondent as "oddly sited"	NR

NR Not recorded. M Male F Female N neutered

*Diagram supplied by respondent (not reproduced in this paper)
‡Perpetrator also physically abused other animals (cases included in Munro and Thrusfield 2001b)

specifically (for example, case 2: "Owner claimed dog sexually abused by lodger, who pled guilty...") In others, respondents had categorized the abuse as physical, but they were subsequently re-classified by the authors as sexual abuse cases because they involved injury to the genitalia or anorectal areas (for example, case 25, in which four equidistant radial splits were present in the anus).

Careful consideration was given to the inclusion of cases 5 and 11. Although there was no direct physical evidence in case 5, the fact that the owner specifically sought examination on the grounds that she suspected sexual abuse, merited documentation. In case 11, the respondent had reported multiple internal hemorrhages around the uterus and within the abdominal wall, but the vulval injury justified its inclusion.

Reasons for Suspicion

The reasons for suspecting sexual abuse varied, encompassing one or more of the following:

- the type of injury;
- the behavior of the owner;
- statements from witnesses;
- admission by the perpetrator.

Thus, types of injury included "Piece of broom handle recovered from distal vagina..." (case 12); behavior of the owner was recorded in one case (case 25) as "...the owner was angry on questioning"); a witness statement was "Dog had been seen being sexually abused" (case 16); and a perpetrator "...admitted abuse" (case 13).

Cases in Dogs

The female dogs included five cases (cases 1 to 5) of known or suspected penile penetrative (vaginal) abuse (three with vaginal injuries), two reports of uterine/vaginal/vulval injury of unexplained origin (cases 6 and 11), six cases involving insertion of a foreign body into the vagina (cases 7, 8, 9, 10, 12 and 13), and one case of insertion of a foreign body into the rectum (case 14). One dog (case 7) with a vaginal foreign body (a knife) died, as did the dog (case 14) with the rectal foreign body (a broomstick).

Cases in male dogs included three with a ligature applied to the external genitalia (cases 17, 18 and 19), two with anal injuries (cases 15 and 16), one (case 20) with severe wounds to the external genitalia (castration), and one with a penetrating wound to the area around the rectum (case 21).

Cases in Cats

Two cats (one female and one of unspecified gender) had penetrating wounds in the rectal and vulval areas (cases 22 and 23). One female cat had an anal injury (case 25), and another had a vaginal foreign body (case 24). The sole recorded male, a young kitten, had suffered trauma to the external genitalia (case 26). This kitten's injuries were fatal.

Cases in Unspecified Species

One respondent recorded unexplained vaginal injuries and another respondent recorded penetrating perianal wounds, described as "oddly sited" (cases 27 and 28, respectively.)

Discussion

The sexual abuse of animals is not a subject that, even today, is frankly and openly discussed, although it has been documented, and usually condemned, for centuries (Beirne 1997). The cases documented in this paper clearly demonstrate that the sexual abuse of animals does occur, and that physical injury may result, some of which may be very severe.

It is the very fact that the abuse involves either the sexual organs, or the anus and rectum, that classifies the abuse as sexual in nature. In this unexplored veterinary field, much benefit is gained from the extensive experience accumulated by the medical profession, on the subjects of child sexual abuse and sexual offences against adults.

Hobbs, Hanks and Wynne (1999a) point out that "child sexual abuse may present in many ways," and that "Normality on examination is common even if the child has been abused." In the current study, cases 2 and 5 displayed no obvious injury. Thus, it can be speculated that, as with children, animal cases may occur that currently are difficult to recognize, with injuries being minimal or absent.

However, Hobbs and others (1999b) also note that violence does occur in some cases of child sexual abuse, and the consequences may be fatal. Sexual offences against adults may be similarly severe (Gee 1985). Such severity is clearly demonstrated in cases 7 (where a knife wound was found deep in the vagina) and 14 (where a broomstick, inserted in the rectum, reached the liver). Common sense, too, indicates that the degree of damage to the female genitalia following penetration, penile or otherwise (for example, cases 1 and 24) is dependent, to some degree, on the size of the animal.

Hobbs, Hanks and Wynne (1999a) note "deliberately sadistic acts," such as lacerations to the dorsum of the penis, and cuts to the vulva or perianal area of children. The current study includes four cases involving penetrating wounds in the area of the vulva and rectum (cases 21, 22, 23 and 28). Intravaginal implements used in this case series include a candle (case 10), a knitting needle (case 8), a fragment of broom handle (case 12), a piece of stick (case 13) and (possibly) a tampon (case 24).

The clinical findings relating to anal sexual abuse of children are well documented (Bamford and Roberts 1997a,b; Hobbs, Hanks and Wynne 1999a), and include anal fissuring, and anal dilatation. Similar findings are also present in the current case series: a cat with the split anus (case 25), a small dog with unexplained damage to the anal mucosa (case 15), and a dog with the dilated anal ring (case 16). Fissuring can occur for other reasons (notably constipation), too. Some authors (for example, Hobbs, Hanks and Wynne 1999a) suggest that differentiation of causes may be possible in children (for example, multiple fissures, or deep fissures, extending onto the perianal skin, in the absence of a history of constipation, identified in some child abuse cases). However, it may be unwise to extrapolate specific features, identified in cases of child abuse to small animals, where there currently is a paucity of recorded cases.

Textbooks on veterinary obstetrics and gynecology do not include sexual abuse in the differential diagnosis of vaginal lesions. Given that some respondents in the current study have reported vaginal injuries caused by sexual abuse (cases 1, 3 and 4), that another respondent reported suspicious "gross vaginal injuries" (case 6), and yet another listed unexplained vaginal injuries (case 27), it seems advisable to add sexual abuse, in appropriate circumstances, to the list of differential diagnoses for vaginal lesions. It *must* be emphasized that—as with NAI—the *circumstances* of the suspected offense must be carefully considered (Munro and Thrusfield 2001a,b)

It is possible that some might at first consider the application of ligatures to the penis or scrotum (cases 17–19) as likely to be a result of an ignorant childish prank. However, Hobbs, Hanks and Wynne (1999a), in their list of more common genital injuries found in abused boys, include a circumferential mark due to a penile ligature. They also note that "It is apparent that doctors have been slow to acknowledge genital injury, and even now fail to name the injury; for example a ligature around the base of the penis may be explained away as punishment for bed-wetting, or aggression on behalf of the abuser." Although veterinarians are not generally involved with the wider issues of abuse in families, the question has to be asked as to why the idea of applying a ligature to these animals occurred to the perpetrator in the first place.

The range of injuries in abused animals reported here mirrors, by and large, the spectrum identified in human victims. However, some types of animal sexual abuse (for example, the use of male animals as active sexual "partners" of women) clearly would not be reported in the current study, although they are known to occur (Adams 1998).

The cases reported are not an accurate measure of the incidence of animal sexual abuse in the UK, because the main aim of the study was to obtain only a basic measure of the extent of abuse in terms of the proportion of veterinary practices that have experienced it (Munro and Thrusfield 2001a). However, the cases do comprise 6 per cent of the 448 cases listed by the 404 veterinarians who completed randomly distributed questionnaires. This suggests that the problem is not trivial.

Beirne (1997), in an essay entitled "Rethinking bestiality," describes the "curious silence" that surrounds the sexual abuse of animals. The same author points out that although "all known societies have likely applied some form of censure to human–animal sexual relations," the abuse has never been censured on the grounds of the harm that it inflicts on animals. It is hoped that this paper will end the silence by stimulating open and frank discussion of the subject.

Acknowledgements

We thank the RSPCA for financial support for the study, the Scottish SPCA for support and interest, and the BSAVA for the provision of members' address labels. Special thanks are also due to Dave Stone and Rhona Muirhead of Edinburgh University for their invaluable help with the study's database. Finally, we are most grateful indeed to the veterinarians who completed and returned questionnaires.

References

Adams, C. J. 1998. Bringing peace home: a feminist philosophical perspective on the abuse of women, children, and pet animals. In *Cruelty to Animals and Interpersonal Violence*, 318–339, ed. R. Lockwood and F. R. Ascione. Purdue: Purdue University Press.

Bamford, F. and Roberts, R. 1997a. Child Sexual Abuse - I. In *ABC of Child Abuse*, 3rd edn., 38–39, ed. R. Meadow. London: BMJ Publishing Group.

Bamford, F. and Roberts, R. 1997b. Child Sexual Abuse—II. In *ABC of Child Abuse*, 3rd edn., 40–46, ed. R. Meadow. London: BMJ Publishing Group.

Beirne, P. 1997. Rethinking bestiality: Towards a concept of interspecies sexual assault. *Theoretical Criminology* 1: 317–340.

Gee, D. J. and Watson, A. A. 1991. Sexual offences. In *Lecture Notes on Forensic Medicine*, 5th edn., 83–88. Oxford: Blackwell Scientific Publications.

Gee, D. J. 1985. Sexual offences. In *The Essentials of Forensic Medicine*, 4th edn., 480–495, ed. C. J. Polson, D. J. Gee and B. Knight, Oxford: Pergamon Press.

Hanks, P. ed. 1986. *Collins Dictionary of the English Language,* 2nd edn. London and Glasgow: Collins, pp. 146, 1769.

Hobbs, C. J., Hanks, H. G. I. and Wynne, J. M. 1999a. Clinical aspects of sexual abuse. In *Child Abuse and Neglect: A Clinician's Handbook*, 2nd edn., 191–271. London: Churchill Livingstone.

Hobbs, C. J., Hanks, H. G. I. and Wynne, J. M. 1999b. Fatal child abuse. In *Child Abuse and Neglect: A Clinician's Handbook*, 2nd edn., 435–446, London: Churchill Livingstone.

Kempe, C. H, Silverman, F. N., Steele, B. F., Droegemuller, W. and Silver H. K. 1962. The Battered-Child Syndrome. *Journal of the American Medical Association* 181: 17–24

Meadow, R. 1997. Epidemiology. In *ABC of Child Abuse*, 3rd edn, 1–4, ed. R. Meadow. London: BMJ Publishing Group.

Munro, H. M. C. and Thrusfield, M. 2001a. 'Battered pets': features that raise suspicion of non-accidental injury. *Journal of Small Animal Practice* 42: 218–226.

Munro, H. M. C. and Thrusfield, M. 2001b. 'Battered pets': injuries found in dogs and cats. *Journal of Small Animal Practice* 42: 279–290.

Tannahill, R. 1989. *Sex in History,* revised edn. London: Sphere Books.

Vermeulen, H. and Odendaal, J. S. J. 1993. Proposed typology of animal abuse. *Anthrozoös* 6: 248–257.

Is zoophilia a sexual orientation? A study

Hani Miletski

Bethesda, Maryland, USA

Abstract

Based on Francoeur's (1991) discussion about the three interrelated aspects of sexual orientation: affectional orientation, sexual fantasy orientation, and erotic orientation, the current study examined the question "Is there a sexual orientation toward animals?" The study involved a 350-item, 23-page, anonymous questionnaire, which was self-administered and returned by postal mail by 82 men and 11 women who had had sexual relations with animals. It was found that some people (the majority of the participants in the current study) have feelings of love and affection for their animals, have sexual fantasies about them, and admit they are sexually attracted to animals—three components that describe sexual orientation. The current study further reveals that the majority of its participants reported being happy and not wanting to stop having sex with animals.

Keywords: *bestiality, mental health providers, sexual fantasies, sexual orientation, zoophilia*

Kinsey, Pomeroy and Martin's (1948) notorious study of the sexual behaviors of 5,300 American men revealed that one man in about 13 engaged in bestiality—sexual contact with animals. Although the majority of these men were farm boys who only had sex with animals on several occasions, the study also included men in their 50s who had sexual relations with animals, and even one man who was over 80 years old (Kinsey, Pomeroy and Martin1948). In 1953, Kinsey et al. found that about five percent of the 5,792 American women in their study engaged in bestiality, as well. The Hunt study (1974), which analyzed data from 982 men and 1,044 women who responded to sex information questionnaires, found the incidence of bestiality to be 4.9% for men and 1.9% for women.

Although these three studies provide evidence that bestiality exists, they are outdated and limited in their findings, since they did not focus on the issue of bestiality. Scientific studies on the motivations for engaging in bestiality, and studies describing the sexual, social, and mental health profile of individuals involved, have been scarce. Peretti and Rowan's study

Address for correspondence: 6917 Arlington, Rd., Suite 202, Bethesda, Maryland 20814, USA. Ph: 301-951-6592; fax: 301-309-0446; e-mail: Hani@DrMiletski.com.

(1983) of 27 men and 24 women, did focus on "chronic bestiality," and attempted to find what helps people to sustain their involvement in bestiality. The study, however, was limited to six pre-determined variables, and the sample size too small for generalization. Donofrio's more recent study (1996) focused on zoophilia—sexual and emotional attraction to animals. However, the small number of "zoo" (as they like to call themselves) participants (8) limits his findings, as well.

In addition, there seems to be a major lack of knowledge among mental health care professionals in regard to treatment options for zoophiles. The renowned Krafft-Ebing (1935) instructed a patient "to be on his guard against masturbation and bestiality, and to seek more the society of ladies." He further "prescribed anaphrodisiacs, advised frugality, slight hydrotherapy, plenty of open-air exercise, (and) steady occupation" (p. 569). Cerrone (1991) suggested family therapy, social assertiveness training, and sex education. The latter was provided to reduce the client's "misperceived thoughts of sex and to educate (him) to the norms of sexual development" (p. 37). McNally and Lukach (1994) recommended a six-month behavioral treatment program comprised of masturbatory satiation, covert sensitization, and stimulus control procedures. Other psychotherapists have reportedly tried to force their zoophile clients to stop having sex with animals (Miletski 1999). Clients have been "locked up" in mental institutions for observations, and treated with drugs and electroshock therapy (Miletski 2001).

I have found the major void of knowledge regarding bestiality/zoophilia alarming. In my opinion, clinical sexologists and psychotherapists need to be equipped with a carefully researched base of knowledge in order to understand the phenomenon and the individuals involved. As professionals helping those who struggle with, and experience ego-dystonic feelings about bestiality and zoophilia, it is important that this research be available and on-going.

Dekkers (1994) and Fox (1994) reported that there are people who are sexually attracted to animals, to the point of preferring animals as sex partners to humans. The term, coined by the "zoo" community on the Internet, "zoosexuality," implies a sexual orientation toward animals (Fox 1994; Tanka 1995; Shepherd 1996; Stasya 1996). And Donofrio (1996) reported that the concept of zoophilia being a sexual orientation was supported by his doctoral study. He therefore suggested using a scale resembling Kinsey's sexual orientation scale, where those who have no interest whatsoever in sexual contact with animals would appear at the zero point of the scale, and those individuals whose sole sexual outlet and attraction was animals would be assigned a score of 6. Along that continuum, between these two extremes, would be individuals who include animal sexual contact in their fantasy, or have had incidental experiences with animals, have had more than incidental contact with animals, place their sexual activity with animals equal to that involving humans, prefer animal contact but engage in more than incidental contact with humans, and those who engage primarily in contact with animals, with only incidental human sexual contact (Donofrio 1996).

I therefore conceptualized the basic research question to be: "Is there a sexual orientation toward non-human animals?" The definition of "sexual orientation" was adapted from Francoeur (1991) in his discussion of homosexuality, heterosexuality, and bisexuality. According to this definition, sexual orientation consists of three interrelated aspects: (1) affectional orientation—who or what we bond with emotionally; (2) sexual fantasy orientation—with whom or what we fantasize having sex; and (3) erotic orientation—with whom or what we prefer to have sex.

Methods

Participants and Procedure

Finding participants for the study turned out to be easier than expected (Miletski 2002). The majority of participants (68, 73%) heard about the study through the Internet, either by seeing a posting or through an Internet friend. Others heard about the study through various advertisements in professional and non-professional publications. More than 160 people contacted me about the study and every participant was required to make telephone or personal contact. This was important as I needed the opportunity to screen the participant for authenticity. Only individuals who had had actual experiences of sexual relations with animals were accepted in the study (many people who only fantasized about this behavior had to be eliminated from the study). I also wanted to know if I was talking to a woman, a man, or a child, and had better chances knowing this by talking to them on the telephone rather than through the Internet. Individuals under the age of consent were eliminated from the study.

The participants were required to provide their postal address, not an e-mail address, as I wanted some control over who received the questionnaire and who returned it. On the Internet people could have forwarded the questionnaire to others who never contacted me, and so I requested the postal address in an attempt to eliminate this possibility. Also, every questionnaire had my original signature on it. If it were to come back photocopied, the questionnaire would have been eliminated from the study.

Every participant received a packet consisting of a questionnaire, a self-addressed, stamped envelope, and a letter of explanation, which was also the consent form. The letter/consent form followed the relevant guidelines set forth by the revised edition of the *Code Of Federal Regulations* (21 CFR 50.20 & 50.25) (1995), which explains the requirements for informed consent of human subjects. Therefore the letter explained the purpose of the study, described the possible discomforts that the participant might experience as a result of filling out the questionnaire, discussed confidentiality and anonymity, pointed out whom to contact for answers to pertinent questions, and stressed that participation in the study was voluntary. The participants were requested to sign the bottom of the letter with their initials or any name they might choose, and to send it back together with the completed questionnaire. Questionnaires that were returned without a signature were eliminated from the study.

The study was conducted as a double-blind study. I was the only person who had access to the participants' addresses, which were destroyed after the questionnaires were sent out. The completed questionnaires were sent back without a return address (for the most part), so that I had no way of knowing who were the respondents.

On September 25, 1996, 25 questionnaires were sent out to 25 volunteers who were randomly chosen from the sample. By January 17, 1997, 15 completed questionnaires came back. These completed questionnaires comprised a pilot study, which was conducted for the purpose of assessing the face and content validity of the questionnaire. The data from this study were incorporated within the final results of the study.

During the month of March 1997, 125 additional questionnaires were sent out to all the volunteers who were not included in the pilot study. The total number of questionnaires which was sent out was therefore 150. Four envelopes were "returned to sender," two participants did not sign the consent form and thus were eliminated from the sample, and one questionnaire was returned blank. Several potential participants called in the months that followed, complaining that they had not received their questionnaire, which was probably lost in the mail.

Of the total 150 questionnaires that were sent out, 93 (62%) participants sent their completed questionnaires back, which comprised this study. Out of 132 men, 82 (62%), sent back their questionnaires, and out of 18 women, 11 (61%) sent back theirs. The study is based on these 82 men and 11 women.

Questionnaire

The questionnaire consisted of 350 items over 23 pages. It included both open-ended questions and questions that required completion or checking off the best answer. This was done for the purpose of obtaining quantitative as well as qualitative information. Some questions involved various items that were measured on a Likert-type scale. The Likert scale was somewhat different for every question, to eliminate a response-set. There was also repetition of some questions with different wording to ensure reliability. The questionnaire included three sections: one section asked for general information and demographics (most of the questions in that section comprised the control variables); another section asked about the participant's psycho-sexual general development; and the third section covered information about sexual relations with animals. The last question simply stated: "Is there anything else you would like to share?"

Results and Discussion

Demographics and Psycho-Social Information

Age

The participants in the study were older than the stereotypical farm, adolescent boy of the Kinsey, Pomeroy and Martin (1948) study. The average age of the men was 38 years, ranging from 19 to 78 years (*SD* = 13.65, median = 37, mode = 21). Five men were 21 years of age, one man was 70 and another was 78 years old. The average age of the women was 36 years, ranging from 21 to 48 years (*SD* = 9.17, median = 35, mode = 47).

Place of Residency

Seventy-one men (87%) and all the women were from the United States. Four men were from Germany, three were from Canada, another three were from the United Kingdom, and one participant was from Australia. The majority of the men (65, 79%) and all but one woman (91%) reported they were connected to the Internet at the time of the study (1997).

Religion

More than half the participants (54, 59%) reported they never attended religious services. Yet, seven men (9%) and one woman reported attending religious services approximately once a month, and another six men (7%) and one woman attended religious services at least once a week. Among these men and women, only two men and one woman reported they no longer engaged in bestiality. These data seem surprising since no current religion condones human–animal sexual contact; in fact, most religions condemn such behaviors.

Education

Concurring with Kinsey. Pomeroy and Martin (1948) and Kinsey et al. (1953), who concluded that people who have sex with animals are often more educated than the general population, in the current study, almost half of the participants (44, 47%) were college graduates or above. One man in the College Graduate category was a graduate of a seminary.

Marital/Relationship Status

Twenty-one men (26%) and three women (27%) reported they were never married or lived in a sexual relationship with another person for more than a month. Almost half the men (37, 45%) and the majority of women (7, 64%) were single at the time of the study. Twenty-six men (32%) and one woman (9%) were married, and ten men (12%) and two women (18%) were divorced (Table 1). Under the category "Other," four men chose to describe their marital/relationship status as it applied to their relationship with their animal lover. One man wrote: "(I am) married to my female horse," and another wrote: "(I am) in an intimate relationship with my animal lover." Moreover, at the time of the study, 21 men and one woman reported being married, living with their spouse, and having sex with animals at the same time.

Other findings in the study show that some of the participants started and/or were having sex with animals at the time of the study, partially because they had no other available sex partners, they were too shy to have sex with humans, they were lonely, and/or they suffered from lack of social interaction. The majority of participants, however, reported this was not the case for them.

Table 1. The participants' marital/relationship status at the time of the study.

Marital/Relationship Status	Men (82)	Women (11)
Single	37(45%)	7(64%)
Married	26(32%)	1(9%)
Divorced	10(12%)	2(18%)
Intimate relationship with a human male lover	6(7%)	0
Intimate relationship with a human female lover	6(7%)	0
Other	5(6%)	0
Living with a human female lover	3(4%)	0
Living with a human male lover	2(2%)	1(9%)
Separated	0	2(18%)
Widowed	1(1%)	0

Mental Health

Menninger (1951) proposed that sexual relations with animals that persists in adult life suggests an "inconclusiveness or unsatisfactoriness in reality," thus the need of a "deviant" love object. Ramsis (1969) suggested that most people who engage in bestiality share a common theme of sexual unhappiness. In the current study, more than half of the participants (47, 57%) had been in psychotherapy, with (an average of) more than two different psychotherapists. It should be noted, however, that this information by itself does not provide any data about the participants' mental nor happiness status; many people, nowadays, go to therapy for a variety of reasons.

Since many consumers of mental health care don't receive a formal diagnosis from their psychotherapists, only 20 participants were able to report they had been diagnosed with mental health disorders. Most of them were depressed, three men were diagnosed with Attention Deficit Disorder, another three men reported they had a "nervous breakdown," two were diagnosed with Bipolar Disorder, and two men were diagnosed with paraphilic disorders.

Surprisingly, only about half of the participants (23, 49%) who had been in psychotherapy told their psychotherapists they have had sex with animals. Almost half (11, 48%) reported they experienced negative reactions from their psychotherapists. Particularly noteworthy, some participants reported their therapists lacked knowledge of zoophilia, laughed upon hearing their client was having sex with animals, or tried to force their clients to stop having sex with animals.

Suicide Attempts

Nineteen participants (23%) reported they tried to commit suicide, on average, more than twice. Another nine noted they had thought about it. It is common knowledge that suicide rates are high among gays and lesbians. They tend to grow up feeling different, lonely, isolated, and unable to talk to others about their homosexual feelings. Since zoophiles have similar experiences, and if zoophilia/zoosexuality is a form of sexual orientation, it may not be surprising that 19 participants in this current study reported they tried to commit suicide, and nine others reported they had thought about it.

Level of Happiness

During the 12 months prior to the study, however, the majority of men (57, 69%) and women (9, 82%) reported they were pretty happy with their personal life (Table 2).

The participants' contentment with life was also evident when, in response to a question, at least 78 participants (94%) reported they did not

Table 2. The level of happiness reported by respondents.

Level of Happiness	Men (82)	Women (11)
Extremely happy	13(16%)	1(9%)
Very happy most of the time	17(20%)	5(45%)
Generally satisfied, pleased	27(33%)	3(27%)
Sometimes fairly unhappy	20(24%)	1
Unhappy most of the time	5(6%)	1

want to stop having sex with animals. The following is a response given by one of the participants to the above question:

> "Definitely not! I enjoy it very much, so do the animals. Neither my friends or my employers think less of me because of it, and I do not believe it to be unethical. It can be a bit frustrating that the majority of society has yet to emerge from the ethical Dark Ages and still believe that we are somehow nasty individuals, but I believe in my own ethical code sufficiently that I have no desire to give up something that I enjoy so much, and which does no harm to anyone else. I can live an enjoyable life, have many friends and enjoy success in my work, without having to stop enjoying animal sex." (Miletski 2002).

The current study revealed an abundance of data about the life, behaviors, values, thoughts, and feelings of its participants, which are beyond the scope of this paper. For a better understanding of the phenomena of bestiality and zoophilia, the reader is invited to read Miletski's recent book (2002).

Is There a Sexual Orientation Toward Animals?

Based on Francoeur's definition of sexual orientation (described earlier), the participants in the current study were asked a variety of questions, throughout the questionnaire, in an attempt to find out whether their relationships with animals corresponded to the three components of sexual orientation.

The Dependent Variable—Sexual Relations with Animals

All the participants reported they had had sex with animals. Eighty-four participants (93%) reported they "enjoyed it very much." None of the participants reported he/she "did not like it" or "hated it."

Fourteen men (17%) and two women reported they had not had sexual relations with animals in the year prior to the study, which means that 68 men (83%) and nine women (82%) had sex with animals during that year. Of the14 men who did not have sex with animals, five indicated they did not have sex with animals in the previous year only because of situations beyond their control—not because they did not want to have sex with animals. In other words, nine men (11%) in the current study had completely stopped having sex with animals. Four of them explained that for them sex with animals was merely something they did during adolescence, and one man reported that sex with animals was a way of acting out for him, before he was "cured."

The Existence of Affectional Orientation Toward Animals

Forty-nine men (60%) and six women (67%) reported the statement "I began having sex with animals because I wanted to express love or affection to the animal" was "completely true" or "mostly true" for them. This was "reason number three" the men (after sexual attraction and curiosity) and "number two" the women (after sexual attraction) provided as to why they initiated sexual relations with animals. Eighteen men (22%) and one woman (11%), reported this sentence was "not true" for them.

Fifty men (74%) and six women (67%) reported the sentence "I am currently having sex with animals because I want to express love or affection to the animal" was "completely true" or "mostly true" for them. This was the "number two reason" (after sexual attraction) for both men and women, which they provided as to why they currently had sexual relations with animals. Six men (9%) and one woman (11%) reported this sentence was "not true" for them.

Seventy-one men and 10 women responded to a question about how old were they when they first realized they were psychologically/emotionally attracted to animals. The purpose of this question was to see if the participants would admit they were psychologically/emotionally attracted to animals. Indeed, 71 men (87%) and 10 women (91%) related that they were psychologically/emotionally attracted to animals. Of the 11 men who did not respond to this question, eight commented that they had never been psychologically/emotionally attracted to animals (the men realized they were psychologically/emotionally attracted to animals at an average age of 11.5 years, and the women at an average age of 8.4 years).

When asked if they would allow other people to have sex with their animal(s), only 16 men (23%) and four women (40%) reported they would not allow it. It is interesting to note, however, that seven men (10%) and three women (30%) reported they would not allow other people to have sex with their animals, as they regarded their animals as mates. Another seven men in this category described love and/or jealous feelings for their animals.

Analysis of the responses of the 76 men and 11 women responding to a question which asked the participants to explain why they defined themselves as "bestialists" or "zoophiles," revealed only eight men (11%) in the sample who appeared to qualify for the definition "bestialist only," as they seemed to have had sex with animals only for the sake of sex. The majority of men (59, 78%) and women (8, 73%) were both "bestialists" and "zoophiles/zoosexuals," since, as some of the participants noted, "a person who has sexual relations with an animal is a bestialist by definition," and "a person who has a love of animals is a zoophile (or zoosexual), from the dictionary definition (zoos = animals + philos = love)."

Obviously, this categorization was subjectively created by myself. But more than the actual numbers, what matters is the way the participants explained their views and described their feelings toward their animals. Expressions such as: "I define myself as a zoophile because my relationships with animals are about love and trust—sex is merely an expression of that love," clearly define the individuals' feelings. Even if the numbers are wrong and there are more "bestialists only" and less "zoophiles/zoosexuals," the participants' reports are loud and clear: there were some participants who seemed to be in love with their animals, and there were others who seemed to have no emotional attraction to animals.

The Existence of Sexual Fantasy about Having Sex with Animals

The majority of men (57, 70%) and women (7, 64%) reported that sexual fantasies about having sex with animals contributed, on some level, to their first sexual encounter with an animal. This means that these participants were fantasizing about having sex with animals before they actually had sex with animals.

The majority of men (61, 76%) reported they "primarily" or "always" fantasized about having sex with animals, at the time of the study. This fantasy was far more popular than any other sexual fantasy reported by the men. Only four men reported they "never" fantasized about it (three of them no longer had sexual relations with animals at the time of the study), and one man reported he "rarely" fantasized about having sex with animals. In comparison, 19 men (24%) never fantasized about having sex with a woman, and 36 men (45%) never fantasized about having sex with a man.

Almost half the women (5, 45%), too, fantasized "primarily" or "always" about having sex with animals, at the time of the study, while the only other popular sexual fantasies mentioned by the women involved watching other humans have sex with animals (40%), and having sex with a man (27%). None of the women reported "never" fantasizing about having sex with animals, and only one woman said she "rarely" fantasized about it.

Again, the participants' reports undoubtedly articulate their fantasy life: the majority of participants fantasized about having sex with animals, and only a few did not. More importantly, the majority of participants began fantasizing about having sex with animals before they had had their first sexual encounter with an animal.

The Existence of Erotic Orientation Toward Animals

All but four men (which makes for 78 men), and all 11 women reported being attracted to certain animals, when asked "To what animals are you most attracted?" In another question, 62 men (76%) and seven women (70%) reported the sentence "I began having sex with animals because I was sexually attracted to the animal" was "completely true" or "mostly true" for them. This, in fact, was the "number one reason" the participants provided for this question. Only eight men (10%) and one woman (10%) said this sentence was "not true" for them.

Sixty-two men (91%) and nine women (100%) reported the sentence "I am currently having sex with animals because I am sexually attracted to the animal" was "completely true" or "mostly true" for them. Again, this was the "number one reason" the participants provided for this question. Only three men (4%) and none of the women said this sentence was "not true" for them.

Seventy-eight men (95%) and 10 women (91%) admitted they were sexually attracted to animals when asked how old were they when they first realized they were sexually attracted to animals. The purpose of this question was to see if the participants would admit they were sexually attracted to animals, and all but four men and one woman reported they were (both, the men and women realized they were sexually attracted to animals at an average age of 13 years).

Nineteen men (27%) and three women (38%) reported believing that bestiality was not a perversion because for them it was the natural thing to do and/or it was like a sexual orientation. And, as mentioned before, 78 (94%) of the participants reported they did not want to stop having sex with animals, mostly because zoophilia was part of who they were, and they liked it.

It is also interesting to note that nine men commented (throughout the questionnaire) about their attraction to animals' pheromones. For example, one of them related: "I enjoy stimuli that are not often found in human sexual relationships; for example, I am highly turned on by olfactory stimuli, and humans by convention rarely allow themselves to have any natural human aroma. Artificial perfumes leave me completely cold, as do conventional standards of 'attractiveness' and 'beauty.' There is something altogether more straightforward and earthy in the experience of animal sex, and it is that which I seek in my sexual activities." Another man commented that his "order of sexual preference is consistent with the sexual appeal of a species' sexual fragrance." This is an example of sexual/chemical attraction on a very basic/biological level.

As mentioned before, analysis of the responses of the 76 men and 11 women responding to a question which asked the participants to explain why they defined themselves as "bestialists" or "zoophiles," revealed only eight men (11%) in the sample who appeared to qualify for the definition "bestialist only," since they seemed to have had sex with animals only for the sake of sex—not because they were sexually attracted to the animals. The majority of men (59, 78%) and women (8, 73%) were both "bestialists" and "zoophiles/zoosexual."

Again, this categorization was subjectively created by myself, and the numbers are less relevant than the way the participants explained their views and described their feelings toward their animals. Expressions such as: "While I have lust for a large number of animals, it is the relationship formed with the animal that is the important part," reinforce the notion of a sexual attraction toward animals. The participants' reports revealed that there were people who had sexual feelings toward animals, while others had none.

One man, in response to the question: "How would your life be different if you had a close intimate/sexual relationship with a human being?" related the following: "...I am zoo exclusive and the very thought of having sex with a human disgusts me. Ask a homosexual if he wants to have sex with someone of the opposite gender."

Eighty-one men and all 11 women rated themselves on a Kinsey-like scale, describing the participants' sexual inclinations toward humans vs. animals. The participants were asked to take into consideration both actual sexual behavior and fantasy. More than half of the men (48, 58%) perceived themselves as having more sexual inclinations toward animals than humans (between 4 and 6 on the scale—Table 3). The majority of the women (9, 82%), however, perceived themselves as being sexually inclined to both humans and animals (between 2 and 4—Table 3).

The findings of the above question, supported by other related questions, clearly indicate that different people have different levels of sexual inclination toward animals. The current study shows that some people (the majority of the participants in the current study) have feelings of love and affection for their animals, have sexual fantasies about them, and admit they are sexually attracted to them. "Is there a sexual orientation toward non-human animals?"—yes, so it appears.

Table 3. The sexual inclinations of the respondents.

Sexual Inclination	Men (82)	Women (11)
0=Exclusively with human beings	2(2%)	0
1=Only incidental animal sex	6(7%)	1(9%)
2=Both animal sex and human sex, but more human sex	14(17%)	4(36%)
3=Equally animal and human sex	12(15%)	3(27%)
4=Both animal sex and human sex, but more animal sex	15(19%)	2(18%)
5=Only incidental human sex	22(27%)	1(9%)
6=Exclusively animal sex	10(12%)	0

Limitations of the Current Study

Unfortunately, the participants' responses could not be statistically analyzed and compared with a "non-zoo" population, as the current study did not include a control group. Comparing the results of the current study with other studies was not a satisfactory option, as studies of "non-zoos" do not ask questions about the participants' feelings, fantasies, and sexual attraction to animals. In addition, the sample size (too small) and the fact that the participants were not randomly chosen for the study do not allow for any meaningful statistical analysis. This lack of statistical analysis renders the study merely a descriptive one (which was the original intent). It is therefore important to remember that the findings of this study cannot be generalized to other "zoos" or bestialists. This study merely describes the lives and some of the behaviors of its 93 participants, as it was intended to be an exploratory, descriptive study.

The majority of subjects (73%) found out about the study through the Internet, which means they were sophisticated enough to use both a personal computer and the Internet. The "zoos" on the Internet may have been more open about their sexual behaviors with animals, about sexuality in general, and about participating in this study as a result of exposure to the Internet and to their peers. It is important to keep in mind that if the sample had been made of more people outside the Internet, the results might have been different.

Other than a short telephone conversation with the volunteers for the study to verify authenticity, and making sure they sent back the original questionnaires, I had no guarantees the subjects were who they said they were, or that their answers were genuine. Although the questionnaires included some repeated questions with different wording to ensure reliability, there is still a chance that participants may have been lying or under- or over-reporting in some of their answers.

Another limitation of the study was the use of open-ended questions, which were primarily analyzed and categorized by myself. As often happens in situations like this, sometimes it was difficult to understand what exactly the participants wanted to convey in their responses.

When it came to the structured questions in the questionnaire, the participants' answers were obviously subjective, and therefore may not have been accurate. For example, when the participants were asked to define themselves in terms of being a "bestialist" and/or a "zoophile," it appeared that they perceived themselves and the various definitions in different ways (making it very confusing). Analysis of their explanations revealed different results from what the participants reported.

Conclusions

The current study did not provide a prevalence rate for people who have this sexual orientation, nor did it provide the causes for having such a sexual orientation. However, the current study very clearly shows that some people (the majority of the participants in the current study) have feelings of love and affection for their animals, have sexual fantasies about them, and admit they are sexually attracted to animals—three components that describe sexual orientation.

Sexual orientation, as we know it, can be fluid, and changes with time and circumstances. People are not "black or white." We can place people on all levels of the Kinsey scale, even when we apply this scale to sexual orientation toward animals. It is logical to assume that the majority of the human race will be placed around the zero point of this Kinsey-like scale (sexual inclination exclusively with human beings), but the current study shows that there are some humans whose place on this scale is definitely not zero. In fact, there are some (probably very few) individuals whose place on this scale would be the other extreme (6 = sexual inclination exclusively with animals).

Moreover, the current study—albeit with its inherent flaws, and the inability to generalize its results, or even to be considered significant—did provide some important insights into the lives of 82 men and 11 women who had sexual relations with animals. One important finding was the fact that the majority of the participants in the study reported being happy and not wanting to stop having sex with animals. In many ways, this study was a breakthrough, as nothing like this had ever been done before and most of the data were new information that could begin to fill the void sexologists have been experiencing about the phenomena of bestiality and zoophilia.

The current study has already opened doors for other studies such as Beetz's (2002) and Williams and Weinberg's (2003), and hopefully, other researchers will follow their path. Future studies should incorporate a larger, random sample, with a control group. A face-to-face interview rather than a self-administered questionnaire, may assure both the participants' understanding of the questions, and the researchers' understanding of the responses in a more objective manner. Standardized tests, such as the Minnesota Multiphasic Personality Inventory (MMPI) could be employed in future studies, where data from the general population is available/known, so that comparison between the general population and people who have had sex with animals could be accomplished.

Lastly, but most importantly, although the majority of participants in the current study reported being pretty happy with their personal lives, zoophiles and bestialists may come to the attention of mental health providers for a variety of reasons (Miletski 2001). They may be dealing with having to live a life of secrecy filled with fear of being outed, anxiety, stress, guilt, shame, low self-esteem, depression, anger, grieving the loss of animal sex partners, coming out, or even wanting to stop having sex with animals. Whether they feel their bestiality is ego-dystonic or they want to work on other issues, it is the mental health provider's professional responsibility to be prepared to assist them. In order to be able to successfully work with zoophiles and bestialists, it is essential to learn more about the phenomena of bestiality and zoophilia and to be prepared to be non-judgmental, open-minded, accepting (which is not the same as condoning), and confidential. Otherwise, zoophiles and bestialists will not reveal their true selves, and this secrecy may negatively impact the therapeutic milieu and progress.

Acknowledgements

There are too many individuals I should thank for helping me with my study than this space allows. Most of all, I would like to thank all the men and women who volunteered and took the time to call me, to fill out a lengthy questionnaire, and to send it back to me. I couldn't have done it without them. I would also like to thank my husband, Arthur Goldberg, my colleague, Cathleen Glass, and the two reviewers for their useful suggestions and for helping me edit this article. Lastly, I would like to take this opportunity to thank Andrea Beetz, Ph.D., for inviting me to write this article, and to share my knowledge/study with my colleagues through this distinguished journal.

References

Beetz, A. 2002. *Love, Violence, and Sexuality in Relationship between Humans and Animals.* Doctoral dissertation. Aachen, Germany: Shaker Verlag.

Cerrone, G. H. 1991. Zoophilia in a rural population: Two case studies. *Journal of Rural Community Psychology* 12(1): 29–39.

Code Of Federal Regulations (21 CFR 50.20 & 50.25). Rev. edn. 1995. Washington, DC: The Office of the Federal Register, National Archives and Records Administration.

Dekkers, M. 1994. *Dearest Pet: On Bestiality.* 2nd edn. Translated by P. Vincent. New York: Verso.

Donofrio, R. 1996. Human/animal sexual contact: A descriptive-exploratory study. Doctoral dissertation. The Institute for Advanced Study of Human Sexuality, San Francisco, CA

Fox, R. 1994. *The alt.sex.bestiality Frequently Asked Questions* [online].

Francoeur, R. T. 1991. *Becoming a Sexual Person.* 2nd edn. New York: Macmillan Publishing Company.

Hunt, M. 1974. *Sexual Behavior in the 1970s.* Chicago, IL: Playboy Press.

Kinsey, A. C., Pomeroy, W. B. and Martin, C. E. 1948. *Sexual Behavior in the Human Male.* Philadelphia, PA: W. B. Saunders Company.

Kinsey, A. C., Pomeroy, W. B., Martin, C. E. and Gebhard, P. H. 1953. *Sexual Behavior in the Human Female.* Philadelphia, PA: W. B. Saunders Company.

Krafft-Ebing, R. V. 1935. *Psychopathia Sexualis.* Rev. edn. Brooklyn, NY: Physicians and Surgeons Book Company.

McNally, R. J. and Lukach, B. M. 1994. Behavioral treatment of zoophilic exhibitionism. *Journal of Behavioral Therapy and Experiential Psychiatry* 22(4): 281–284.

Menninger, K. A. 1951. Totemic aspects of contemporary attitudes toward animals. In *Psychoanalysis and Culture: Essays in Honor of Geza Roheim.* ed. G. B. Wilbur and W. Guensterberger. New York, NY: International Universities Press.

Miletski, H. 1999. Bestiality/zoophilia—An exploratory study. Doctoral dissertation. The Institute for Advanced Study of Human Sexuality, San Francisco, CA.

Miletski, H. 2001. Zoophilia—implications for therapy. *Journal of Sex Education and Therapy* 26(2): 85–89.

Miletski, H. 2002. *Understanding Bestiality and Zoophilia*. Bethesda, MD: East West Publishing, LLC.

Peretti, P. O. and Rowan, M. 1983. Zoophilia: factors related to its sustained practice. *Panminerva Medica* 25: 127–131.

Ramsis, J. A. 1969. *Casebook: Animal Love.* Pendulum Books, Inc.

Shepherd, R. 1996. Fur of the beast: The life and loves of a dog lover. Unpublished manuscript.

Stasya. 1996. *Stasya's Home Page* [online]. Http://www.av.qnet.com~stasya.

Tanka, S. 1995. *Shuunka's View of Zoophilia and Zoosexuality* [online]. Shuunka@net-com.com.

Williams, C. J. and Weinberg, M. S. 2003. Zoophilia in men: a study of sexual interest in animals. *Archives of Sexual Behaviors* 32(6): 523–535.

New insights into bestiality and zoophilia

Andrea M. Beetz

Department of Veterinary Medicine, University of Cambridge, UK

Abstract

Sexual contact with animals, usually referred to as bestiality, occurs in many different forms and is practiced with different underlying motives. In particular zoophilia, where an emotional bond to the animal plays a key role besides the sexual aspect, has received more attention during the past few years. This article reports primarily on three recent studies—Beetz (2002), Miletski (2002), and Williams and Weinberg (2003)—in which the authors investigated bestiality and zoophilia among large samples of people, providing new insights into these phenomena. Their findings on the species involved in sexual contacts with humans, the practices engaged in, the development of sexual interest in animals, the personality of men practicing sex with animals, links with mental health problems, the differences between zoophilia and bestiality, and the role of the Internet are reported. Furthermore, the attitudes of philosophers and society towards the practice of bestiality are discussed.

Keywords: *bestiality, personality, society, zoophilia*

In a time when sexuality is discussed openly in most Western societies, sexual contact with animals remains one of the last taboos and is rarely addressed. If it is mentioned, the general population usually reacts with ridicule, disgust, moral outrage, and sometimes also indifference or voyeuristic curiosity, but rarely with informed comment and/or a desire for more knowledge about this phenomenon. Bestiality is the commonly used term to refer to sexual contact with animals, although definitions vary and some do not include all sexual acts with animals: some only include penetration of or by the animal. Little information is available about its prevalence today. While several researchers have collected data on bestiality as one behavior among many others in the field of sexual abuse or interpersonal violence (see Beetz, pp. 145–169 in this issue of *Anthrozoös*), only a few studies have involved exclusive investigations of bestiality, and even fewer have acquired a large sample size. In the present article, the findings of recent studies by Beetz (2002), Miletski (2002), and

Address for correspondence: Dr A. M. Beetz, Schillerstrasse 28a, 91054 Erlangen, Germany. Ph: +49 09131-4000 455; e-mail: andrea.m.beetz@gmx.de.

Williams and Weinberg (2003), utilizing large, voluntary samples of people admitting to engaging in sexual practices with animals, will be reported in more detail. Their data support the existence of another phenomenon closely related to bestiality—zoophilia—where the key feature, in addition to sexual interactions, is a strong emotional involvement with the animal.

Only during the last five years has there been a surge of scholarly interest in sexual contacts with animals, possibly due to the findings of recent research; practitioners and scientists in the mental health professions as well as animal protection, law, criminology, and anthrozoology have started to discuss bestiality and zoophilia in more detail. In this paper, I hope to promote further discussion by presenting a summary of the most recent research on sexual contacts with animals.

Firstly, an overview of the variety of forms of sexual contact with animals that occur and the different definitions of bestiality and zoophilia will be given. Secondly, a short description of the studies of Beetz (2002), Miletski (2002), and Williams and Weinberg (2003) will follow to serve as the primary source of information on the following aspects of sexual contacts with animals: the species involved in sexual contacts with humans, the sexual practices engaged in, the development of a sexual interest in animals, the personality of men practicing sex with animals, links with mental health problems, a comparison of zoophilia and bestiality, and the importance of the Internet. Thirdly, the views on bestiality taken by philosophers and society in general will be addressed. Finally, some conclusions about the relevance of bestiality and zoophilia to different professions and society will be drawn.

Types of Sexual Contact with Animals

In former times, bestiality and other forms of "deviant" sexual behavior such as homosexuality and anal intercourse were subsumed under the term "sodomy" (Stayton 1994). Today, this term is not used in relation to sexual contact with animals any more. Now, "zoophilia" is the term preferred by many of the people who engage in sex with animals, and they call themselves "zoophiles" or "zoos." Zoophilia is also the term most often employed by clinicians. It was introduced into the field of sex research by Krafft-Ebing in 1894 (Schmidt 1969), who described several different forms of sexual contact with animals in his well-known work *Psychopathia Sexualis*. While he labeled all non-pathological sexual contact with animals as "bestiality," cases that were like an animal fetishism were defined as "zoophilia erotica," and pathological cases as "zooerasty." In contrast to this, Masters' (1962) definition of zooerasty focuses on a lack of emotional involvement in the sexual act with animals, which made it comparable to masturbation. Karpman (1954), however, addressed only "sexual excitement experiences with stroking or fondling of animals" (p. 15) as "zoophilia," and used the terms "bestiality" and "zooerasty" when referring to any sexual act between humans and animals. Several other authors employed the term "zoophilia" to name an exclusive or predominant desire for sexual acts with animals (Masters 1966). For Money (1986), zoophilia represented "a paraphilia of the stigmatic/eligibilic type, in which sexuoerotic arousal and facilitation or attainment of orgasm are responsive to and dependent upon engaging in cross-species sexual activities…" (Money 1986, p. 273). It is important to mention, too, that sometimes a general love of animals *without* any sexual interest has been called zoophilia. And in the non-scientific literature, the term "Egyptian" can be found sometimes in reference to sexual practices with animals, in the same way that "French" is used to refer to oral practices, and "Greek" to homosexual contact (Bryant 1977).

In mental health, zoophilia is listed among the paraphilias. Its definition in the third edition of the *Diagnostic and Statistical Manual of Mental Diseases* (DSM) states: "The act or fantasy of engaging in sexual activity with animals is repeatedly preferred or the exclusive method of achieving sexual excitement" (American Psychiatric Association <APA> 1980, p. 270). The fourth edition, DSM-IV (APA 1994), named the following, more exact, criteria for paraphilias "…recurrent, intense sexually arousing fantasies, sexual urges, or behaviors generally involving 1) nonhuman objects, 2) the suffering or humiliation of oneself or one's partner, or 3) children or other nonconsenting persons, that occur over a period of at least six months (Criterion A). … The behavior, sexual urges, or fantasies cause clinically significant distress or impairment in social, occupational, or other important areas of functioning (Criterion B)" (APA 1994, p. 523). Also, legal consequences may be regarded as a clinically significant impairment. Thus, in states where bestiality is illegal, this criterion usually applies.

Other phenomena closely related to sexual contact with animals have been reported by several authors. A case of "zoomimic masochism"—the human abasement to an animal state, for example, playing an animal in a sadomasochistic context—was described by Hirschfeld (1956). Stekel (1952), however, named the identification with an animal for the purpose of sexual performance, for example, playing the role of a dog or performing as a rooster in a costume, "zoanthropo-sexual infantilism." Later, Bornemann (1990 cited in Rosenbauer 1997) defined the mimicking of animals in general as "zooanthropy," and the mimicking of animals in a mainly sexual context as "zoomimic."

Money (1986) reported a highly specific form of zoophilia, "formicophilia." In cases of formicophilia, sexual arousal and achievement of orgasm is caused by small creatures such as snails, frogs, insects, or ants "creeping, crawling, or nibbling the genitalia and perianal area, and the nipples" (Money 1986, p. 76). Further, Money (1986) claimed bestiality to be an animal fetishism.

McNally and Lukach (1991) described a case of "zoophilic exhibitionism," where a mildly mentally impaired man masturbated in front of large, male and female dogs, but never exposed himself to women. The man desired and practiced no sexual contact with the dogs other than frottage: rubbing himself on them, and letting the animals lick off his ejaculate. In addition, he engaged in "zoophilic voyeurism"—peeping through windows to watch dogs fornicate. Even though this patient had satisfactory sexual relationships with women, he preferred zoophilic exhibitionism. Rosenbauer (1997) used the terms "mixoscopia bestialis" and "mixoscopic zoophilia" to refer to sexual arousal derived from observing animals mating.

Just as "sadism" describes the deriving of sexual pleasure from inflicting pain or harm or causing death in an interpersonal sexual context, "zoosadism" relates to the experience of sexual pleasure when torturing or killing animals, sometimes in combination with sexual practices. Although there are a variety of cases involving different kinds of animals and types of injury, in particular the sexual abuse of poultry (such as chicken, ducks, or geese), and rabbits should be noted. The penetration of the animal in combination with strangling the animal or breaking its neck provides sexual stimulation to the zoosadist; also there is physical stimulation due to the spasms of the dying animal (Beetz 2002; Miletski 2002).

More recently, the following definitions of zoophilia and bestiality were put forward by Miletski (1999) and Kurrelggyre (1995 cited in Miletski 2002). According to these authors, zoophilia is characterized by an emotional attachment to animals that causes a person to prefer an animal as a sexual partner, or includes a sexual attraction. Bestiality, however, describes any sexual contact between humans and animals, or any physical contact with animals that leads to sexual excitement and pleasure for the person involved. Miletski (2002) stated that bestiality and zoophilia do not represent distinct categories, but rather may occur in combination, or flow into each other over a spectrum of human–animal relations. These definitions of bestiality and zoophilia are also used in the same way by most zoophiles themselves. Furthermore, Miletski (2002) proposed in reference to her data that some cases of zoophilia even fulfill the criteria for a sexual orientation (Francoeur 1991 cited in Miletski 2002) towards animals, named "zoosexuality." "Zoosexuals" have an emotional as well as a sexual attraction to, and relationship with, animals (Miletski 2002). As with other sexual orientations such as homosexuality or bisexuality, it is difficult to distinguish one sexual orientation from another, as sometimes fantasy life and desire are not translated accordingly into actual behavior. Several studies (e.g., Beetz 2002; Miletski 2002; Williams and Weinberg 2003) have documented that the majority of persons engaging in sex with animals also have sexual experiences with human partners, and that some prefer to have sex with both animals and humans. Only for people who have a predominant or exclusive attraction to animals and who do not practice sex with humans would the diagnosis of an exclusively zoosexual orientation apply.

The information cited above demonstrates that the types of sexual interest in animals vary widely. It has been claimed (Massen 1994) that a latent sexual interest in animals can be found in many people—an indicator of this might be the frequently observed interest and sexual excitement people have watching the mating of animals (Massen 1994). At the other end of the continuum, the sexual orientation towards animals is located (Miletski 2002). Massen (1994, p. 57) proposed that there were nine basic forms of zoophilia, and added that often several of these forms occur in combination:

1) incidental experience and latent zoophilia,
2) zoophile voyeurism (also called mixoscopic zoophilia),
3) frottage (Massen 1994 described this as physical contact as source of pleasure),
4) the animal as a tool for masturbatory activities,
5) the animal as a surrogate object for a behavioral fetishism (such as sadomasochistic practices, sexual murder, etc.),
6) the animal as fetish (fixation on one specific kind, breed, or individual),
7) physical contact and affection,
8) the animal as a surrogate for a human sex partner, and
9) the animal as deliberately and voluntarily chosen sex partner.

This list is not exhaustive in its description of types and motivations, and especially zoophilia and zoosexuality need to be investigated further.

With regard to actual sexual practices involving animals, nearly every practice found in a human–human sexual context is possible (for an overview see Beetz 2002). Voyeurism, exhibitionism, masturbatory practices (rubbing against the animal for own sexual stimulation or masturbating the animal), oral–oral contact, oral–genital contact (with animal or human receiving), and vaginal or anal penetration of the animal or the human can be found, sometimes in combination with masochistic or sadistic practices, or in a context of interpersonal violence and physical and sexual abuse. No accurate data on the prevalence of the different practices and contexts exist. However, the studies of Beetz (2002), Miletski (2002), and Williams and Weinberg (2003) give details on the preferences for certain practices and species among people who admit to having sexual contact with animals.

Recent Studies on Sexual Contacts with Animals

Methodologies

Even though other authors have also contributed a lot of valuable information on bestiality and zoophilia, this paper concentrates mainly on three studies: Beetz (2002), Miletski (2002), and Williams and Weinberg (2003). Their studies are special not only because they collected data from large samples (about 100 people), concentrated exclusively on bestiality and zoophilia, and have been conducted during the last five years, but also because they worked with completely voluntary samples and utilized the Internet to make initial contact. A similar approach with volunteers was employed by Peretti and Rowan (1983), but their study focused on factors related to the *sustained* practice of bestiality, named chronic zoophilia.

In this paper, mainly the results from the Miletski (2002), Beetz (2002) and Williams and Weinberg (2003) studies are reported; however, where relevant, the results of Peretti and Rowan (1983) will be referred to. A short introduction to the methodologies and sample characteristics of each study will now be provided.

Peretti and Rowan (1983) obtained data from 27 men and 24 women, aged 17 to 28 years, who had engaged in sexual contact with animals at least twice a month for a minimum of two years. All participants were referred to the researchers by their physicians and volunteered to be interviewed face-to-face about factors related to their sustained practice of bestiality. All of them also had satisfactory human sexual relations.

Miletski's (2002) research began with her dissertational project. She collected material from conventional sources as well as the Internet, and developed a 350-item questionnaire which asked for personal data and information on childhood history, sexual history, current sexual behaviors and preferences for humans and animals, and animal ownership. Miletski established contact with participants mainly via the Internet, although she had also tried other ways, such as through advertisements in a newspaper. Questionnaires were posted after personal contact had been made by phone, as Miletski wanted to have some control over who received the questionnaire. Overall, data from 82 men and 11 women who had engaged in sexual contact with animals were obtained. Thirty-six percent were between 19 and 29 years old, 27% between 30 and 39, and 36% between 40 and 49.

More than 90% were Caucasian, and about half of the sample were college graduates or people with a higher qualification. All of the women and 87% of the men came from the United States; 5% came from Germany. Twenty-six percent of the men had never been married or had lived in a sexual relationship with another person for a month or more.

At around the same time, Williams and Weinberg (2003) used an approach similar to Miletski's, starting data collection in 1999. They had established contact with people practicing sex with animals via a specialized website, and had designed a questionnaire that volunteers could answer online, after contacting the authors via email. The questionnaire asked about "shared identity"—how participants labeled themselves with respect to their sexual interest in animals and how they related to others with the same interest—the nature of their sexual interest in animals, sexual contact with animals, human sexual desires and contacts, and the balance of animal and human sexual desires (p. 526). Data from 114 men were analyzed; only five women and one transgender responded to the questionnaire. Ninety-one percent of the male participants, all of whom were White, lived in the United States. Their ages ranged from 18 to 70 years, with a median age of 27 years. Sixty-four percent had never been married and were single, and 83% had at least some college education or had completed college. Thirty-four percent of the men were living in a rural area and 36% in a large city or its suburbs.

In 2000, Beetz (2002) collected data from 113 men and three women who reported sexual experiences with animals and had volunteered to answer a set of questionnaires. Similar to the other studies, one questionnaire sought information on personal data, animal ownership, sexual experiences with humans and animals, preferences in relation to sex with animals and humans, and mental problems. As the purpose of the study was to also collect data on the personality of people engaging in sex with animals, the following standardized instruments were used: California Psychological Inventory (CPI, Gough 1987); Feelings, Reactions and Beliefs Survey (FRBS, Cartwright and Mori 1988; Höger 1994, 1995, personal communication 1999); Inventory of Interpersonal Problems (IIP, Horowitz et al. 1994, 2000); Attitude and Preference Questionnaire (Zuckerman 1979; Tellegen 1982; Levenson et al. 1995; Möller, personal communication 1999); and the Thematic Apperception Test (TAT based on Murray 1943), rated for motive content. These allowed for the comparison of this group to normal populations in relation to personality and motives. All questionnaires, with the exception of the CPI, were made available in English and German, as first contacts on the Internet had shown that a large percentage of the possible participants were from German-speaking countries. The assessment of personality was an important part of this study. It provided information that has never been obtained before and addressed some of the prejudices about bestiality, such as that persons who engage in bestiality are more violent, are psychopathic, are less socially competent, or have more interpersonal problems than those who don't engage in it. Participants were approached in German- and English-speaking Internet chatrooms and via the main German and American mailing-lists dealing with zoophilia/bestiality. Volunteers had to contact the author via email, and the questionnaires were sent and returned online. In addition to the questionnaires, detailed interviews were conducted with 36 men and three women, 22 of them in a face-to-face situation. At interview, questions were repeated from the questionnaire in more detail, and participants were asked further about their developmental history of sexual experiences and relationships with humans and animals. Due to the small number of female participants, only the data of the men will be reported here (for information on the female participants, see Beetz 2002). Beetz's sample differs from those of Miletski (2002) and Williams and Weinberg (2003) in far fewer came from the US: only 35% of the men were from the US; 32% came from Germany. Overall, in Beetz's (2002) study, 40% of the participants came from the US, Canada or Australia, and 56% from Europe; thirty-seven percent of the respondents answered the German version of the questionnaires. The average age of the men was 30 years (*SD* = 9.88) and about 70% had at least some college education or a higher level of education. At the time of data collection, only 21% of the men lived in a stable relationship with a human partner.

Important to note is that all of the authors cited above had personal contact (face-to-face) with their participants either at gatherings and/or during their data collection. Such contact can provide a more comprehensive impression and understanding of the situation, compared with just questionnaire data.

Summary of Results

In the following paragraphs, the findings of Beetz (2002), Miletski (2002), and Williams and Weinberg (2003) on species involved, practices engaged in, motives involved in sexual contacts with animals, the development of zoophilia, sexual relations with human partners, the personality of zoophiles, links with mental health problems, the distinction between bestiality and zoophilia, and further diverse findings are reported. It needs to be kept in mind that the reported data cannot be generalized to the whole population of people engaging in sex with animals, and that they only refer to the voluntary samples investigated in these studies, or a similar population. Even though this is a limitation of the findings, the data can give some

Table 1. Percentage of men who had sexual contact with different types of animals, and the sex of animal involved.

	Source		
	Mileteski (2002)	**Beetz (2002)**	**Williams and Weinberg (2003)**
Dogs/other canines:	—	69.9%	63%
Male dogs:	90%	60.2%	—
Female dogs:	72%	46.0%	—
Horses/other equines:	—	50.4%	37%
Male horses:	54%	35.4%	—
Female horses:	52%	40.7%	—
Cattle/other bovines:	—	7.1%	—
Male cattle:	18%	3.5%	—
Female cattle:	40%	6.2%	—

insight into this special sub-group of people. It is very likely, though, that the data suffer from positive selection bias (i.e., people who were low on violence and placed more importance on emotional involvement), due to only volunteers participating.

Species and Sex of Animals Involved

Table 1 shows the different animals (and sex of) which male participants reported to have had sexual relations with. It seems surprising that dogs are the species most often selected for sexual relations and that male dogs are preferred, as the common assumption is that usually female farm animals are most commonly approached. Fewer participants, but still about one in every two, had ever had sex with a horse. A remarkable difference comes up in relation to sexual experiences with cows and bulls. In Miletski's (2002) sample, far more men reported sexual relations with these animals than among the men investigated by Beetz (2002); overall, only 14% of Beetz's sample indicated having sexual experiences with farm animals. One explanation may be related to the percentage of European/German participants in Beetz's sample; possibly in Germany it is more difficult to gain access to cows, as in many parts of the country these animals are mainly kept in closed stables. Remarkable, too, is the percentage of men approaching bulls in a sexual manner, a practice that would be rather dangerous.

Other animal species approached by Miletski's (2002) participants were female sheep (21%), female felines (20%), male felines (17%), female swine (16%), female goats (13%), and male swine, male goats and female fowls

(10% each). A few participants also reported sexual interactions with llamas/camels, donkeys, deer, tapirs, rabbits, rhesus macaques, wolves, large cats (e.g., lions and tigers), and a rhinoceros. Beetz (2002) found that two men in her sample had sexual experiences with large cats and that five men wished to have sexual encounters with large cats. In addition, four men were sexually interested in dolphins and two men had had actual sexual experiences with this species (intercourse did not happen in all cases). Even though this might sound rather unbelievable, my impression from the personal interviews I conducted (Beetz 2002) is that these reports are true (Beetz 2002).

Miletski (2002) found that the men in her sample had sex with a number of animals of one species. The average number of dogs involved in sex with one man was 22, ranging from 1 to 400. Williams and Weinberg (2003) reported an average of eight animals involved in sex with one man. With regard to the sex of the animals involved, no clear preference was observed. More than half of the men in Beetz's (2002) study had sexual experiences with both male and female dogs and horses.

Sexual Practices Involving Animals

Table 2 shows the sexual practices that participants engaged in with animals. Masturbating an animal was a very common practice reported by the participants, mostly performed with male dogs. Also, vaginal intercourse with female animals, in particular horses, occurred quite frequently. Not only dogs but also horses were reported to perform fellatio, and a large proportion of men orally stimulated the animal. Anal penetration of the animal was practiced less often, but definitely more often with horses than with dogs. The high percentage of men who encouraged anal penetration by the animal is surprising, especially when you consider the huge risk of injury to men who allow horses to penetrate them. Further practices reported were voyeurism in relation to animals mating or human–animal sexual performances, French kissing, fisting the animal, frottage, and sexual "play" with the animal's urine and feces.

Among Beetz's (2002) sample, 52.6% of the men who were sexually active with dogs reported sexual contact with this species several times a week, and 18.6% did so about once a month. Only 21.4% of the men who were active with horses indicated sexual contact several times a week; the majority had sex with horses between once a month and once a year. Miletski (2002) found that among her sample the average frequency of sexual animal contact was 2.96 times per week.

About half of Miletski's (2002) sample, 48%, admitted to at least once having used force on an animal in relation to sex, with over half of them

Table 2. Sexual practices that the respondents engaged in with animals.

	Source		
	Miletski (2002)*	**Beetz (2002)****	**Williams and Weinberg (2003)****
Sexual Activity	**Female, Male**		
Masturbating animal	38%, 64%	—	—
Dog	—	96.2%	—
Horse	—	87.5%	—
Receiving oral sex	14%, 23%	—	—
Dog	—	78.2%	44%
Horse:	—	30.4%	14%
Performing oral sex	34%, 42%	—	—
Dog	—	79.5%	—
Horse	—	66.1%	—
Performing vaginal intercourse	55%,—	—	—
Dog	—	50.0%	74%
Horse	—	69.6%	100%
Performing anal intercourse	5%, 8%	—	—
Dog	—	14.1%	24%
Horse	—	39.3%	50%
Receiving anal intercourse	—, 34%	—	—
Dog	—	64.1%	63%
Horse	—	17.9%	32%

* the first number relates to men who "always or primarily engage in this behavior" with female animals, while the second relates to male animals. The species of animal was not differentiated in this study.

** the percentages relate to men who engage in sexual relations with this species. The sex of the animal was not differentiated in these studies.

stating that they had only done it in the past. About 10% of Beetz's (2002) sample admitted to the use of force; six men (5.3%) reported that they had harmed/injured an animal by engaging in sex with it, and three of them stated that this was not intentional—it had been an accident. Three men investigated by Mileltski (2002) reported that they had at least on one occasion been forced by others to engage in sex with animals.

Motives/Reasons for Sexual Contact with Animals

A variety of reasons or motives for engaging in sexual relations with animals were found. Among Miletski's (2002) sample, the reason reported by most men, 91%, was "sexual attraction," followed by the "wish to express love and affection to the animals" (74%). A reason for 67% of the men was that "animals are accepting and easy to please," and 66% claimed that "the animal wants it." Further reasons were "relieving sexual tension" (40%), "I can only trust animals" (39%), the wish to "experience something different" (25%), "I identify with the animal of my gender" (24%), "I see it in pornography" (21%), "loneliness" (15%), "no human partner" (12%), "too shy to have sex with humans" (7%), and "If I did to humans what I do to animals, I would be arrested" (3%).

In Beetz's (2002) study, the following percentages of men indicated different reasons for their sexual involvement with animals: "it is innate" (57.5%), "it is learned" (17.7%), "permanent contact with animals" (28.3%), "lack of other sexual outlet" (12.4%), "opportunity" (24.8%), "by accident" (18.6%), "animal-sex is less complicated" (26.5%), and "animals are better lovers" (30.1%).

In Williams and Weinberg's (2003) study, the reasons most often reported by male zoophiles as having had "a lot or more than little" influence on their sexual interest in animals were "sex with animals is pleasurable" (73%), "a desire for affection" (49%), "family had a household pet" (23%), "not being popular" (19%), "unpleasant sexual experiences with humans" (14%), and the "fear of AIDS or other sex diseases" (7%).

The data of Peretti and Rowan (1983) show that the men in their sample engaged in *chronic* zoophilia for a number of reasons: "sexual expressiveness" (93%), "sexual fantasy" (81%), "no need for negotiation" (74%), "no human social involvements necessary" (63%), "economical reasons" (59%) and "emotional involvement" (26%).

As these data show, the lack of a human partner is a minor reason for engaging in bestiality.

Development of Sexual Activities with Animals

First sexual experiences with animals occurred predominantly in the early or mid-teens. Williams and Weinberg (2003) found the onset of zoosexual activity to be between 11 and 14 years of age, and two-thirds of their sample had sex with an animal before the age of 17 years. Of the men investigated by Beetz (2002), two-thirds had their first sexual experience with animals by the age of 17, and about 50% had it between the ages of 12 and 15 years; surprisingly, 6% had their first sexual contact with animals under the age of 10 years. Miletski (2002) presented similar data: the average age for men experiencing their first zoosexual encounter was 13 years. As with current sexual contact with animals, the species involved during first sexual contact with animals varied widely. In Miletski's (2002) sample, 35% of men had their first sexual experience with their own pet, and 38% with the animal of someone they knew. Twenty-nine percent had their first sexual contact with an animal in a home environment, and 44% had it outside. Only 12% had negative feelings after their first experience, while 30% had mixed feelings, and 58% had positive feelings.

First sexual fantasies with animals started on average around the same time as the first experiences: at the age of 13 years (Beetz 2002). And although about half of the men interviewed (n = 36) in Beetz's (2002) study said they had a normal relationship with animals in childhood, about one-third claimed to have had a much closer attachment to their pets than other people have.

Only a small amount of the collected data can be reported here; further information about the development of sexual activities with animals can be found in Miletski (2002) and Beetz (2002). The case descriptions in their reports certainly provide a thorough insight into the diversity of life histories leading to sexual contact with animals.

Sexual Activities with Human Partners

Sexual activities with human partners were quite common among the participants in the studies reviewed here. The experience of heterosexual intercourse was reported by 83% of the men surveyed by Miletski (2002), and about two-thirds had had at least some homosexual encounters. Sexual experiences with both sexes were also common (43%) in Beetz's (2002) study. However, 17% of Williams and Weinberg's (2003) sample and 24% of Beetz's (2002) sample never had any sexual experiences with human partners (an association with lower age was found by Beetz 2002). Also, 13% of the men investigated by Williams and Weinberg (2003) and 24.8% of the men investigated by Beetz (2002) indicated that they were not sexually interested in either men or women. In relation to a self-assigned sexual orientation, Beetz (2002) found that 44% of here participants regarded themselves as predominantly heterosexual, 15% as bisexual, and 16% as predominantly homosexual. Respectively, the figures from Miletski (2002) are 72%, 8%, and 20% (no category for "not interested in humans" was included).

Even though the majority of men (61.9%) in Beetz's (2002) study would have liked to have had a steady relationship with a human partner, not many did at the time of data collection. Asked about other sexual interests or activities, about 7% indicated an interest in sex with minors/children (Beetz 2002; Miletski 2002), sadistic sex (1.8%), masochistic sex (4.4%), bondage (8.8%), use of feces/urine (11.5%), and exhibitionism (9.7%) (Beetz 2002). Nine percent of Miletski's (2002) participants reported to have forced someone to do something sexual that they didn't want to do.

Looking at the importance of sex with humans versus sex with animals in the whole sex history of the participants, it becomes clear that sex with animals was more important than, or was preferred to, sex with humans for the majority of participants: this was true for more than two-thirds of Williams and Weinberg's (2003) sample and 56.6% of Beetz's (2002). Sex with humans and sex with animals were equally important for about a quarter of the participants in each study, and 14.2% of Beetz's (2002) respondents and just a few in the Williams and Weinberg (2003) study said that sex with humans was more important for them. Interestingly, placing more importance on sex with animals was found to be connected to a lack of sexual experiences with human partners (Beetz 2002; Williams and Weinberg 2003). In relation to actual experiences—not preferences—about a quarter of the male zoophiles (Beetz 2002) stated that animal sex took place rather rarely, while for most (58.4%) sex with animals was the predominant or only kind of sexual activity (apart from masturbation).

Personality of Men Engaging in Sex with Animals

Beetz (2002) compared the personality data of her male sample, using a number of psychometric scales (detailed earlier), with the available normative data. Social desirability was checked via control scales. According to the results, the zoophile sample described themselves as more shy, uneasy in social situations, and more detached and self-sufficient, in comparison to the normal population. They felt more uncomfortable with people and were less open to feelings in human relationships. Even though they enjoyed company, they did not like the investment of effort and time necessary for such contact. They preferred freedom from obligations and were more self-centered and distrusting. Overall, they described themselves as having more difficulties in interpersonal relationships than the normal population (Beetz 2002).

When asked directly about sociability, the majority of male zoophiles reported to have friends and to be of average sociability (Miletski 2002), but the majority (51.3%) in Beetz's (2002) study socialized primarily with other zoophiles. Not unexpectedly, the zoophile men indicated they were more norm-doubting and unconventional. With regard to "sensation seeking," the zoophiles were found to be less susceptible to boredom, while results on experience-seeking and thrill- and adventure-seeking did not show any significant differences from the normal population. On the psychopathy scales, the participants showed the same degree or even fewer signs of primary or secondary psychopathy than the normal population. Furthermore, participants indicated they had average to above-average empathic abilities, were slightly better judges of what and how people feel and think, and were as or more sympathetic and helpful than the normal population (Beetz 2002). One explanation for better developed empathic ability could be their need to perceive and interpret nonverbal communication from animals accurately; this is especially needed if they do indeed approach unfamiliar animals in a non-forceful way, as many claim to do.

The need for control and dominance by the zoophiles studied by Beetz (2002) did not differ from the normal population. A sub-group which reported that they engaged in "fence-hopping" to get access to animals described themselves as more self-confident, self-accepting, and assertive than the rest of the sample, and showed a lower level of social inhibition.

It is important to note that the results detailed above cannot be generalized to all persons engaging in sex with animals, as the time and effort to participate in this study (3- to 4-hour questionnaire, 2-hour interview) most likely led to a biased sample.

Mental Health Problems

Many of the male zoophiles had been in psychotherapy: 50% of Miletski's (2002) sample and 38.1% of Beetz's (2002) sample. It is interesting that some of these men were from Germany, where it is not as common as in the US to seek professional help for mental health problems. Only a few of the men were in treatment because of their sexual interest in animals (7.1%, Beetz 2002); more frequently, depression was the cause (12.4%, Beetz 2002). Other reasons for seeking therapy were social problems (5.3%, Beetz 2002), attention deficit hyperactivity syndrome, phobias, compulsions, antisocial personality disorder, family problems, nervous breakdowns, and paraphilic behaviors.

Of the men in Miletski's (2002) study, 36% were happy most of the time, 33% were generally satisfied, about 30% were rather unhappy, and 22% had tried to commit suicide at least once. Whether their problems were directly linked to their sexual interest in animals remains unclear, but 85% of the men did not want to stop their sexual activity with animals (Miletski 2002). Half of the men in psychotherapy told their therapist about their sexual activity with animals, and in half of those cases reactions from the therapist were negative, ranging from ridicule, threats to report to the police, disbelief, and a lack of knowledge about the existence of such practices, to an attempt at a forceful cure (Miletski 2002).

Bestiality and Zoophilia

As explained earlier, the main difference between bestiality and zoophilia (as defined by zoophiles, Beetz 2002; Miletski 2002) is that zoophilia, besides involving sexual contact with animals, also includes an emotional involvement with the animal. When asked about their emotional involvement, only a few men (3.5%, Beetz 2002) reported having no emotional attachment to the animal they were having sex with, while about 20% indicated a "normal" attachment, like one has with a pet, and the majority (76.1%) reported a very strong emotional attachment, comparable to love between human partners. However, it is obvious that a person engaging in infrequent sex with a variety of animals will not have a strong emotional involvement to each, while this may be different when the person has a relationship with his/her own animal or has regular contact with an animal—participants talked about the strongest involvement they ever had.

Another similarity to human relationships is the phenomenon of "falling in love" with animals: this was reported by 78.8% of the men in Beetz's (2002) study. A further indicator of a strong emotional aspect to sexual relations with animals came out when male zoophiles were asked if they would allow others to have sex with their animal; only 24% said they would generally allow this, 53% would only allow it under certain circumstances to certain people, while 23% would not allow this at all (Miletski 2002). Frequently, jealousy was cited as a reason for not allowing others to have sex with their animal, again pointing to a strong emotional relationship. However, in the study by Beetz (2002), more than 75% of the men stated that they had at least once had sex with another person's animal, without the knowledge of the owner.

When asked to classify themselves as zoophiles or bestialists, half of Miletski's (2002) sample put themselves in the first category, and only 9% in the latter. About one-third stated that they were both bestialists and zoophiles, probably because they had an emotional involvement with one specific animal and also practiced sex with other animals, with whom they had no special emotions.

Animal Protection Involvement and Internet Usage

A further interesting finding, although not unexpected, was that about one-third of the male zoophiles in Beetz's (2002) study reported to be actively involved in animal protection. Less than 10% of the men in the studies by Beetz (2002) and Miletski (2002) worked with animals in their jobs.

As nearly all the participants of the three studies described in this paper were approached via the Internet, the time respondents spent on the Internet, also in terms of socializing via this medium, was of interest. Beetz (2002) found that the average time spent online per week was 25 hours, with a quarter of the sample spending between six and ten hours online and another quarter between eleven and 20 hours; about 9% spent more than 30 hours online and 14% spent more than 40 hours. A large proportion of this online-time was spent talking with other zoophiles privately and/or in a chatroom related to their shared sexual interest. For most of the men interested in sex with animals, the information available on the Internet (FAQs, chatrooms) and the exchange with like-minded others was perceived as helpful and important: many lacked knowledge, and the medium allowed them for the first time to openly talk about their sexual activity with others. Although a lot of sites on the Internet provide pornographic material, and some sites provide instructions on how to have sexual interactions with different species (which could potentially promote the practice of bestiality), the value of the online-exchange cannot be underestimated for persons who struggle with issues of self-acceptance and isolation due to their sexual interest. Of the zoophile men investigated by Beetz (2002), 18% reported serious problems with acceptance by others and society, 6% had trouble finding a partner and were lonely, and 8% had trouble keeping their sexual activity with animals secret. All of these men had already found the resources available on the Internet.

Philosophy and Bestiality

Overall, philosophy has rarely dealt with the issue of bestiality. Kant (1724–1804), in his ethical theory, and in relation to the practice of bestiality, emphasized a person's duty to oneself (Denis 1999). Besides his formula for humanity, his concept of nature's purposes and unnatural uses of a person's sexual capacities addressed deviant sexual practices. According to Kant, bestiality—like masturbation or homosexuality—was not only against the animal nature and humans' natural instincts, but it degraded people "below the level of animals" (cited in Denis 1999, p. 232); by engaging in bestiality, humans neglected their duties to themselves. Kant argued that bestialists should be cast out of human society and be deprived of all human rights (Masters 1962). Arguments such as Kant's can still be found in today's discussions on bestiality, although in Western societies his opinions on masturbation and homosexuality were revised.

Although at the time most philosophers more or less agreed with Kant's view, Jeremy Bentham (1748–1832) approached the discussion on bestiality from a different perspective (Crompton 1978). He pointed to the danger of innocent people being accused of bestiality by those attempting to discredit them, and argued that giving the issue of bestiality too much attention only shocked people and that this was not helpful in any way (Crompton 1978).

That bestiality still remains largely a taboo subject very likely results from an attempt to defend strong ethnic, religious, or institutional regulations (Davies 1982). Rather than considering bestiality as a problem of a single person, the involved animal, and its owner, it has been perceived as a violation of the whole community (Brown 1952, cited in Davies 1982).

Society and Bestiality

The subject of bestiality rarely comes to the attention of the general public, but when it does common reactions include ridicule, disgust, interest or fast dismissal, as it is regarded as being of little importance and of low prevalence. Only in animal protection, animal ethics, and research in the fields of criminology, sociology, mental health and psychology has bestiality been discussed thoughtfully. Whether there is any acceptance for this behavior in society, even in states where bestiality is not criminalized, is not known, as very few people actually convey their opinion on the subject. It seems, though, that there is still a strong influence of old religious and moral values/codes, condemning any kind of sexual contact with animals. Therefore mentioning bestiality frequently evokes rather emotional and extreme reactions. Even when scholarly research in this area is published, authors are faced—depending on the results—with criticism, and not just from their peers. Therefore any statement addressing bestiality needs to be made with caution, respecting the sensitivity of the subject.

The problems attached to addressing the topic of bestiality can be evidenced from a recent situation. Peter Singer, well known for his work *Animal Liberation* (1975), reviewed Midas Dekker's (1994) book on bestiality, *Dearest Pet*, online in *Nerve* magazine (Singer 2001a), and was heavily criticized by many people, among them Piers Beirne, who regards all sexual acts with animals as interspecies sexual assault (Beirne 1997). Supposedly, Singer promoted an "attitude of liberal tolerance towards bestiality" (Beirne 2001, p. 44) and suggested that bestiality should be tolerated as long as it does not involve cruelty (Beirne 2001). It seems Singer did not expect the wave of criticism, especially from the animal rights movement, that followed his review, which he wrote to provoke a more frank discussion and was not meant for a scholarly forum (see Beirne 2001; Singer 2001b).

More surprising were the reactions to a new contribution on bestiality—not in the field of science, but in the arts. The play "The Goat" by Edward Albee, which portrays the life and problems of a man who falls in love with a goat, ran on Broadway in 2004, as well as internationally, and even received an award. Naturally, opinions about it were diverse. However, when I saw the play (in Germany), no openly negative reactions were observed or heard among the audience during the break or after the play had ended. Instead, sympathetic reactions seemed to prevail, influenced undoubtedly by the self-selection of the audience: they chose to be confronted with a different view on sexual contacts with, and love of, animals.

Conclusions

The phenomenon of sexual contact with animals has started to lose its status as a taboo: it is appearing more often in scholarly publications, and the public are being confronted with it, too. Animals today have become an integral part of people's lives, especially in their role as pets, but are also intensively used (e.g., in farming, medical research, sport). Their sexuality is controlled in breeding and farming for economic reasons, and for a large proportion of pets it is strongly influenced by neutering and the deprivation of any outlet for sexual needs. And while nonsexual touch, sharing a bed with a pet, cuddling or kissing it, and emotionally very close relationships are widely accepted, sometimes the boundary between nonsexual and sexual touch becomes blurred, and one may lead to the other. In these cases, the intention of the person or the gain of personal sexual pleasure seem to be important diagnostic criteria. But one could ask provocatively whether the intention of the human really makes a difference for the animal involved, and what kinds of intrusions are actually stressful and harmful. Sexual contact between humans and animals, especially its violent forms, is definitely an issue that needs to be discussed in animal protection. However, it needs to be addressed less emotionally and in perspective with other sexually intrusive acts which are performed on animals and which society supports (e.g., pregnancy testing, artificial insemination).

As the studies in this paper show, more knowledge about the practice of bestiality and zoophilia is needed. In particular, in the mental health professions information and a rational and professional handling of patients who disclose their sexual activity with animals are needed. And the available information on the different forms of bestiality, the reasons for them, and possible links with mental health problems could be helpful to the patient.

Little is known about sexual contact between women and animals, as it is difficult to find women engaging in bestiality who are willing to participate in research. More research in this area is definitely needed. Some studies, even though they were conducted a long time ago, show that bestiality is certainly practiced by women, and at a prevalence that cannot be dismissed.

The Internet has proven to be important for people who have a sexual interest in animals, as well as being a useful place to conduct research in this area. Obviously there are also negative aspects of this online-information, such as animal pornography. It seems, though, as if a large proportion of this pornographic material is not produced for zoophiles, rather it is for consumers who just want to see something extraordinary.

To conclude, sexual contact with animals—in the form of bestiality or zoophilia—needs to be discussed more openly and investigated in more detail by scholars working in disciplines such as animal ethics, animal behavior, anthrozoology, psychology, mental health, sociology, and law.

Acknowledgements

I thank the reviewers for their helpful comments on the first version of this paper.

References

APA. 1980. *Diagnostic and Statistical Manual of Mental Disorders.* 3rd edn. Washington, DC: American Psychiatric Association.

APA. 1994. *Diagnostic and Statistical Manual of Mental Disorders.* 4th edn. Washington, DC: American Psychiatric Association.

Beetz, A. 2002. *Love, Violence, and Sexuality in Relationships between Humans and Animals.* Doctoral dissertation. Aachen, Germany: Shaker Verlag.

Beirne, P. 1997. Rethinking bestiality: towards a concept of interspecies sexual assault. *Theoretical Criminology* 1(3): 317–340.

Beirne, P. 2001. Peter Singer's "Heavy Petting" and the politics of animal sexual assault. *Critical Criminology* 10: 43–55.

Bornemann, E. 1990. *Enzyklopädie der Sexualität.* Frankfurt/Berlin: Ullstein.

Brown, J. S. 1952. A comparative study of deviations from sexual mores. *American Sociological Review* 17(2): 135–146.

Bryant, C. D. 1977. *Sexual Deviancy in Social Context.* New York: New Viewpoints/ Franklin Watts.

Cartwright, D. and Mori, C. 1988. Scales for assessing aspects of the person. *Person-Centered Review* 3: 176–194.

Crompton, L. ed. 1978. Jeremy Bentham's essay on "paederasty", part 2. *Journal of Homosexuality* 4(1): 91–107.

Davies, C. 1982. Sexual taboos and social boundaries. *American Journal of Sociology* 87(5): 1032–1063.

Dekkers, M. 1994. *Dearest Pet: On Bestiality.* New York: Verso.

Denis, L. 1999. Kant on the wrongness of "unnatural" sex. *History of Philosophy Quarterly* 16(2): 225–248.

Francoeur, R. T. 1991. *Becoming a Sexual Person.* 2nd edn. New York: Macmillan Publishing Company.

Gough, H. G. 1987. *CPI-California Psychological Inventory. Administrator's Guide.* Palo Alto, CA: Consulting Psychologists Press, Inc.

Hirschfeld, M. 1956. *Sexual Anomalies.* New York: Emerson Books.

Höger, D. 1994. FRBS—Feelings, Reactions and Beliefs Survey. Desmond S. Cartwright. 2. revidierte Fassung, 1994. Übersetzung A. Angleitner, U. Büsselberg, D. Höger und C. Jennings.

Höger, D. 1995. Deutsche Adaption und erste Validierung des "Feelings, Reactions and Beliefs Survey" (FRBS) von Desmond S. Cartwright. Ein Beitrag zur konzeptorientierten Erfassung von Effekten der Klientenzentrierten Gesprächspsychotherapie. In *Forschung zur Klientenzentrierten Psychotherapie*, 167–183, ed. J. Eckert. Köln: GwG-Verlag.

Horowitz, L. M., Alden, L. E., Wiggins, J. S. and Pincus, A. L. 2000. *IIP - Inventory of Interpersonal Problems.* USA: The Psychological Corporation.

Horowitz, L.M., Strauß, B. and Kordy, H.; unter Mitarbeit von Alden, L. E., Wiggins, J. S. and Pincus, A. L. 1994. IIP-D. *Inventar zur Erfassung Interpersonaler Probleme. Deutsche Version. Manual.* Weinheim: Beltz Test GmbH.

Kant, I. 1991. *Metaphysics of Morals.* Translation by M. J. Gregor. Cambridge: Cambridge University Press.

Karpman, B. 1954. *The Sexual Offender and His Offenses.* Washington DC: Julian Press.

Krafft-Ebing, R. v. 1894. *Psychopathia Sexualis.* 9th edn. Stuttgart: Enke.

Kurrelgyre. 1995. The Zoophilia Report. Accessed at An373492@anon.penet.fi.

Levenson, M. R., Kiehl, K. A. and Fitzpatrick, C. M. 1995. Assessing psychopathic attributes in a non-institutionalized population. *Journal of Personality and Social Psychology* 68(1): 151–158.

Massen, J. 1994. *Zoophilie. Die sexuelle Liebe zu Tieren.* Köln: Pinto Press.

Masters, R. E. L. 1962. *Forbidden Sexual Behavior and Morality.* New York: Julian Publishers.

Masters, R. E. L. 1966. *Sex-Driven People.* Los Angeles, CA: Sherbourne Press, Inc.

McNally, R. J. and Lukach, R. M. 1991. Behavioral treatment of zoophilic exhibitionism. *Journal of Behavioural Therapy & Experimental Psychiatry* 22(4): 281–284.

Miletski, H. 1999. Bestiality/zoophilia: An exploratory study. Doctoral dissertation. Institute for Advanced Study of Human Sexuality, San Francisco, CA.

Miletski, H. 2002. *Understanding Bestiality and Zoophilia.* Bethesda, MD: Author.

Money, J. 1986. *Lovemaps.* New York: Irvington Publishers, Inc.

Peretti, P. O. and Rowan, M. 1983. Zoophilia: factors related to its sustained practice. *Panminerva Medica* 25: 127–131.

Rosenbauer, F. 1997. *Sexueller Kontakt mit Tieren.* Semesterarbeit in der Veranstaltung "Sexuelle Störungen" im SS 1997 bei Prof. C. Möller. Siegen. Accessed at: http://www.rosenbauer.de/ha-zo.htm.

Schmidt, W. 1969. *Neurosenpsychologische Aspekte der Sodomie.* Dissertation. Universität München.

Singer, P. 1975. *Animal Liberation.* New York: Avon.

Singer, P. 2001a. Heavy petting. *Nerve* (March/April). www.nerve.com/Opinions/Singer/heavyPetting.

Singer, P. 2001b. Clarification (of) the circumstances and intent of (my) review of Midas Dekkers' book *Dearest Pet.* Princeton University press release, April 14.

Stayton, W. R. 1994. Sodomy. In *Human Sexuality: An Encyclopedia.* ed. V. L. Bullough and B. Bullough. New York: Garland Publishing.

Stekel, W. 1952. *Patterns of Psychosexual Infantilism.* New York: Liveright.

Tellegen, A. P. 1982. *Manual for the Multidimensional Personality Questionnaire.* Department of Psychiatry, University of Minnesota.

Williams J. W. and Weinberg M. S. 2003. Zoophilia in men: A study of sexual interest in animals. *Archives of Sexual Behavior* 32(6): 523–535.

Zuckerman, M. 1979. *Sensation Seeking: Beyond the Optimal Level for Arousal.* New Jersey: Lawrence Earlbaum Association.

Bestiality: Petting, "humane rape," sexual assault, and the enigma of sexual interactions between humans and non-human animals

Frank R. Ascione

Department of Psychology, Utah State University, USA

"So the man named all the birds and all the animals; but not one of them was a suitable companion to help him." (*Good News Bible* [1978], p. 3)

This collection of papers represents a serious attempt to address a topic whose mention elicits reactions ranging from incredulity and disgust to complex scholarly discourse, discourse that may, on occasion, jeopardize an academic career (Wang 2004). Andrea Beetz, the editor of this special issue, has created an excellent forum in which ideas and research about bestiality are presented in a manner that will, no doubt, prompt further debate and, one hopes, continued research examining sexual interactions between humans and non-human animals—interactions that may sometimes be benign but that may also represent abuse and cruelty shrouded in sexual violence.

In this commentary, I will provide an overview of what I believe are the seminal contributions to the literature offered by these papers, tie research on bestiality to broader issues of human and animal welfare, examine challenges to conducting research on bestiality, and offer areas for future exploration of this important topic. But, before I begin to address these matters, I illustrate the enigmatic nature of the study of bestiality by referring to two recently published articles.

At the benign end of the spectrum, Sandnabba et al. (2003) reported on a study conducted in Finland with 2- to 7-year-old children using the "Day-Care Sexuality Questionnaire" (DCSQ). The stated purpose of the study was to gather information about day-care personnel's reports of children's sexual *behaviors*. Included in the DCSQ are two items listed under the heading "Sex with animals," suggesting an attempt to tap early manifestations of bestiality. However, examining the two items suggests otherwise. Item 101 reads, "Is interested in animals' reproduction" and Item

Address for correspondence: 2810 Old Main Hill, Logan, Utah 84322-2810, USA. E-mail: Frank.Ascione@usu.edu

102, "Talks about the sexual behavior of animals." Although normative information about the developmental course of children's sexual knowledge, interests, and behavior is important in its own right and for assessing deviant developmental patterns, the two items referred to above appear to represent sexual curiosity as distinct from budding paraphilia.

"There is the sweat and smell of it." (Shepard 1996, p. 151)

At the malevolent extreme of the continuum is a case reported by Earls and Lalumière (2002). They describe the history and current status of a 54-year-old man serving a five-year prison sentence for an animal cruelty conviction related to an incident in which the man inserted his arm into a mare's vagina and perforated it; the mare died from the injury. The man's self-reported motivation was that he was emotionally as well as sexually involved with the mare and was jealous of the mare's "interest" in a near-by stallion. This was the fourth of a series of sentences this man received for sexual assaults on horses. He was diagnosed (Diagnostic and Statistical Manual of Mental Disorders—DSM-IV) with bestiality (Paraphilia-Not Otherwise Specified) and Antisocial Personality Disorder and was offered treatment while in prison, but declined. The authors report that the man grew up on a farm, began with sexual activity with chickens, and later focused his sexual activity on mares. Phallometric testing with photographs of men, women, children, dogs, cats, chickens, sheep, cows, and horses resulted in elevated arousal levels only to horses.

The continuum anchored by these two extreme examples is the frame within which these six papers examine bestiality, its history, its definition(s) and manifestations, and the moral and legal judgments about its acceptability that have existed and ought, perhaps, to exist in contemporary society.

> "No studies are available which provide sufficient firmly based material to allow a proper understanding of the subject, which undoubtedly has many facets." (from Shenken's [1964] four-decades-old paper on bestiality, p. 141)

The first paper, by Miletski, is remarkable in its scope, providing an overview, from prehistory to the present, of various cultures' practices involving human–animal sexual activities, one of the most strongly disapproved forms of human–animal interactions according to one recent survey of US adults (Vollum, Buffington-Vollum and Longmire 2004). From representations of bestiality in ancient cave paintings to internet news groups, the author describes a dizzying array of sexual acts humans are believed to perform with animals of every conceivable species, from insects to large mammals and other primates. Both human males and females are reported to engage in bestiality and animals may be oblivious to the sexual acts in which humans involve them, may experience gradations of physical and psychological harm, or may die from injuries inflicted during sexual interactions with humans. Religious condemnations of these practices are described together with cases where bestiality has been considered a part of worship or religious ritual. Bestiality has been described as adolescents', primarily boys', exploration of their emerging sexuality, as a practice enhancing human health and genital development, as a commodity, for example, in pornography and live performances, and as a practice for which animal co-actors have been condemned to death. The modern history of the practice of bestiality awaits further study and analysis, study and analysis that will be able to rely on more direct, and perhaps more reliable and objectively verified, accounts of this phenomenon.

The second paper, by Bolliger and Goetschel, considers bestiality in the context of animal welfare law. Four distinctions offered by the authors are critical to this area of study. First, it is clear we must distinguish between human fondness for animals that may include pleasurable physical contact (e.g., petting, scratching, performing activities related to animal hygiene and health) and cases where animals are the object of human erotic desire and interest. Second, bestiality may involve acts that range from the benign in their physical impact on animals to those that cause serious injury or death. Third, in discussing whether to sanction or condemn bestiality, both human and animal dignity must be considered. Finally, examination of bestiality should distinguish cases where it should be considered a human mental health issue from cases encompassed by criminal law (though the potential overlap between the two is obvious). These distinctions are explored in an excellent overview of past and current legal statutes related to bestiality in various European States, the United Kingdom, and North America. The authors devote considerable attention to the issue of animal "consent" in cases of non-injurious bestiality, raising important questions about how humans might evaluate the "willing" participation of animals, cases where animals are reported to seek human sexual contact, and whether animals' participation is the result of human force, power, or "grooming" (e.g., a pedophile threatening a child victim or slowly introducing a child to more and more intrusive sexual activities). The parallels to the sexual abuse of infants, children and young adolescents, the elderly, and individuals with cognitive or physical disabilities are made clear.

The third paper, by Beetz, introduces a further complication to this discussion but addresses a critical issue. Just as bestiality practices lie on a continuum from benign, non-intrusive acts that may produce no physical harm to those that result in clear physical harm or the death of an animal, human sexual offending against other humans (I will refer to this as "sex offending" to distinguish it from bestiality) lies on a similar continuum of harm to humans. Sex offending ranges from non-touch offenses such as exhibitionism to violent, sadistic rape and sexual murder. The author provides examples of research examining the relations among bestiality, non-sexual cruelty to animals, and sex offending. Here, too, important distinctions are offered as challenges for future research: Are we studying human fantasies about bestiality or actual sex acts with animals? Are we measuring and reporting on individual behaviors or aggregate measures with labels that cloud the nature and severity of the acts? If the issue of "consent" is so critical to our judgment of bestiality, how does this square with clearly "non-consensual" acts that are considered socially acceptable (e.g., slaughtering animals for food)? Should we use a form of triage to decide on which cases of bestiality warrant the attention of mental health professionals and law enforcement or should any form of human–animal sexual contact be subsumed under the monolithic label, "bestiality"?

The fourth paper, by Munro and Thrusfield, reports on the results of a survey of UK veterinarians assessing their encounters with animals, brought to their clinics, who had been sexually abused (this is one of four reports by these authors on various forms of animal maltreatment). The victims, most often dogs and cats, were both immature and adult animals, perpetrators included children, adolescents, and adults who resided with these pets, and the litany of injuries is horrific to read. It is important to note that veterinarians are more likely to see cases where bestiality has resulted in clear physical trauma as distinct from incidents that were less invasive and harmful. Nevertheless, the message is clear: veterinarians will encounter cases where family pets have been sexually abused, abuse so violent at times that animals may die from their injuries. The parallel to child welfare is clear. Pediatricians may encounter children who have been sexually abused and be called on to document the nature of the injuries for legal proceedings.

The fifth paper, by Miletski, explores the issue of whether bestiality should be considered a sexual orientation (akin to heterosexuality, homosexuality, bisexuality—and pedophilia?). Respondents to a survey of individuals who reported having had sexual relations with animals described the practices they engaged in and the nature of their relationships with the animals with whom they had sex. Both men and women completed the survey (which included questions about a variety of demographic variables), with a number (57%) indicating that they had been or were undergoing psychotherapy. Only half of those in therapy indicated that they had discussed their bestiality with their therapists. The reports of those surveyed (perhaps biased by self-selection factors) generally indicated subjective well-being with regard to their bestiality. Many considered animals to be objects of their affection (an awareness many reported as first occurring around puberty), the focus of their fantasies about sexual encounters, and the focus of their erotic attraction and acts; meeting these three criteria suggests to the author that bestiality may, indeed, qualify as a sexual orientation.

The sixth and final paper in the series, by Beetz, reports on a similar survey of individuals who report practicing bestiality and, representing a critically important addition to the literature, the results of various personality assessments allowing comparison of this group's scores with normative data. Personal interviews with some participants in their homes allowed the author to observe the quality of human–animal interactions. Details are provided on the proportion of participants engaging in specified forms of human–animal sexual contact (e.g., masturbating the animals, receiving oral stimulation from the animal) and participants' stated motivations for having sex with animals. For many, bestiality emerged in their early adolescence. Some of the motivations offered by participants suggest animals are preferred as sexual partners, may be considered more convenient than pursuing sexual relations with human partners, or that sex with animals would make contracting HIV-AIDS or other STDs less likely. Again, we are confronted with a continuum of relations, ranging from those based on using the animal as sexual object to those that appear grounded in a deep, affectional relationship. In fact, a third of the participants reported that they were involved in animal protection activities.

A small minority of participants reported that animals had been unintentionally injured during sexual activity. Although a number of differences in personality measures were found between participants' and norm groups' scores, perhaps the most intriguing was that participants scored higher than the norm group on a measure of empathy. The vast majority of participants had no interest in curtailing their practice of bestiality and the author notes that DSM-IV diagnostic criteria require that individuals experience significant emotional distress or impairment in their everyday functioning before certain behaviors are considered a mental disorder.

> "...people, and animals all play out the contradictions we feel in their human-like animality and our animal-like humanity." (Shepard 1996, p. 72)

"It is not altogether surprising, therefore, that this physical and sensual dimension of the human–pet relationship occasionally overflows accepted boundaries." (Serpell 1996, p. 33)

One of the challenges to progress in the study of bestiality will be the need to achieve consensus on the naming and definition of the phenomenon, an issue that continues to be examined in the areas of animal abuse, more generally, and cruelty to animals (Ascione 2005). I agree with Munro and Thrusfield (this issue) that substituting "interspecies sexual assault" raises more questions than it solves—the topic of forced sex between different species of nonhuman animals is an interesting one but takes us far a field from the focus of these essays. It may be advisable to borrow established categories from the child sexual abuse and adult sexual assault literatures in order to standardize data collection. For example, the continuum of humans' sexual abuse of other humans may range from non-touching, non-invasive acts (e.g., exhibitionism, voyeurism) to minimally invasive acts that are unlikely to produce physical harm (e.g., fondling, frottage, oral-oral contact) and, at the extreme, intrusive acts that may cause injuries, may be accompanied by threatened or actual physical violence, and may produce permanent damage or the victim's death (e.g., rape of an immature animal, object rape). Existing inventories of sexual abuse/sexual assault could be easily converted to questions about human–animal sexual interactions and would offer some level of uniformity in assessment across research studies. Consideration should also be given to categorizing bestiality involving dead animals or their body parts (see, for example, Randall, Vance and McCalmont 1990). In any case, future research on people who practice bestiality should routinely include detailed information on the nature and frequency of the sex acts they engage in and the level of physical harm, if any, animals may have suffered.

Research on bestiality must also eventually move beyond convenience samples of self-selected individuals who, in some cases, may participate in research surveys as a way of affirming the acceptability of their own sex practices. This will be especially challenging for social scientists who would like to trace the etiology and developmental course of bestiality (or study rare cases where bestiality may emerge as a side effect of drug therapy—see Jimenez-Jimenez et al. 2002). Parental reports of their children's behavior may be helpful but we must keep in mind that bestiality may be enacted covertly, forcing us to consider how we might ethically ask young people to self-report on their experiences. This may be less of a challenge with samples of clinically distressed children than with normative samples.

It is also clear that greater attention is needed by the veterinary community in establishing guidelines for assessing injuries and other symptoms in animals that would lead to a diagnosis of bestiality. These guidelines would serve clinical needs but, in jurisdictions where bestiality is a crime, may also serve forensic purposes. Experts in human medicine may need to be involved in establishing these guidelines, as in some cases it is the human who is injured by animal sexual acts (Weigand, Schmidt and Kleiber 1999).

In a recent report, Gomes et al. (2000) describe ten cases involving two men and eight boys seen in hospital clinics for injuries to their genitals that were the result of animal bites (eight by dogs and one each by a horse and a donkey). The child victims ranged in age from 5 months to 13 years (also, see Donovan and Kaplan [1989], who raised questions about possible lapses in parental supervision of infants and young children bitten by animals). Although the details of the attacks were not provided, this study raises questions about the behavior of animals that may be involved in bestiality. If an animal has been trained to mouth or lick the genitals of a human who finds these acts pleasurable, how might the animal react if these same acts are attempted with an unfamiliar human who protests and resists? This is not idle speculation, as in one report 58% of a small sample (n = 12) of sadomasochistically oriented men who reported practicing bestiality had children (Sandnabba et al. 2002). No information was provided on whether or not these children were exposed to their fathers' acts of bestiality and, if so, whether children attempted to imitate such parental behavior.

Another question for the veterinary community is the issue of zoonoses that may be related to bestiality. Which human sexually transmitted diseases (STDs) may infect animals in the course of bestiality (Rehan 2003)? Are there animal STDs or other irritants (for example, in canine sperm [Holden and Sherline 1973]) that may affect human health? Answers to these questions may be of special interest in forensic evaluations of children who have been sexually abused by humans or who have engaged in bestiality.

> "I gave no consent…There was a moment when it occurred to me to wonder if bestiality is a sin for the beast, for that was certainly my role in the festivities." (the words of Father Emilio Sandoz in Russell's [1996] novel, *The Sparrow*, pp. 394–395)

The articles on animal welfare law and new insights present the most comprehensive discussions of whether bestiality can, in certain cases, be considered "consensual." To me, this is the most enigmatic element of debates over the acceptability of bestiality. As creatures dependent on human care and nurturance, animals may be similar in status to infants, children, young adolescents, the frail elderly, and individuals with cognitive disabilities when we consider their ability to consent to participation in sex acts. Behavioral cues may be obvious when consent is refused, for example, when the victim protests with screams or displays facial signs of distress or pain but can the absence of such behavioral cues be taken to indicate consent? A man who has intercourse with an intoxicated adult women is considered to have raped her if, prior to her intoxication, she did not verbally consent to intercourse. Should we interpret animals' silence during acts of bestiality as "consent"? What of cases where animals give nonverbal signs suggesting arousal and pleasure? Laws against the sexual abuse of infants, children, and youth are *not* null and void if these young victims experience some level of pleasure during sex acts with adults. And are humans who practice bestiality even more culpable if they use non-aversive operant conditioning techniques to train animals to participate in sex acts? I do not pretend to have answers to these questions, but it is clear that those scholars with greater expertise in this area will eventually need to address them.

Having skirted this difficult ethical and moral dilemma, let me now turn to cases where bestiality is coerced, clearer in its negative impact, and more clearly embedded in serious human mental health issues. The first article by Beetz (this issue) most clearly addresses these cases and I will simply list additional clinical examples.

Duffield, Hassiotis, and Vizard (1998) report on a sample of seven 8- to 16-year-old juvenile sex offenders, evaluated at a clinic, who admitted to having sex with animals (from kissing and touching animals' genitals and mutual masturbation to forced penetration). Some of the features reported in these cases included:

- bestiality co-occurring with physical animal abuse
- an adolescent contracting Toxocara from oro-genital contact with a dog
- forcing a sibling to watch while an adolescent had sex with a dog

Abel, Osborn, and Twigg (1993) reported on a large sample (1,000+) of individuals diagnosed with various paraphilias (fetishism, pedophilia, sadism, rape, etc.). Self-reports of bestiality (undefined) were made by 8.3% of adolescent clients and 8% of adult clients. Hunter et al. (1993) examined the history of ten female, juvenile sex offenders in residential treatment, noting that 20% reported engaging in bestiality. Both Sgroi and Sargent (1993) and Itzin (1998) report cases where children were forced to engage in bestiality.

Finally, I would be remiss, given my professional identity as a developmental psychologist, if I did not close my discussion by focusing on one area that, I believe, has not been given sufficient attention in discussions of bestiality thus far—the exposure of children to adult bestiality. If some adults consider bestiality a benign or even desirable sexual practice and if these adults have children, how will they address these sexual practices with their own offspring (see contrasting cases reported on pages 374 and 377 in Beetz [2002])? Should children be trained in "appropriate" ways of having sex with animals? If children observe one or both parents engaging in bestiality, does this affirm for children the normalcy of these practices? What if children begin to imitate adult bestiality but do so with the pets of their friends and neighbors? Do children who engage in bestiality show a greater likelihood of acting out sexually with other children? The answers to these and related questions await scientific scrutiny and dissemination to professionals in animal welfare, child welfare, law enforcement, and mental health.

References

Abel, G. G., Osborn, C. A. and Twigg, D. A. 1993. Sexual assault through the life span: Adult offenders with juvenile histories. In *The Juvenile Sex Offender*, 104–117, ed. H. E. Barbaree, W. L. Marshall and S. M. Hudson. New York: Guilford Press.

Ascione, F. R. 2005. *Children and Animals: Exploring the Roots of Kindness and Cruelty*. West Lafayette, IN: Purdue University Press.

Beetz, A. 2002. *Love, Violence, and Sexuality in Relationships between Humans and Animals*. Doctoral dissertation. Aachen, Germany: Shaker Verlag.

Donovan, J. F. and Kaplan, W. E. 1989. The therapy of genital trauma by dog bite. *Journal of Urology* 141: 1163–1165.

Duffield, G., Hassiotis, A. and Vizard, E. 1998. Zoophilia in young sexual abusers. *Journal of Forensic Psychiatry* 9: 294–304.

Earls, C. M. and Lalumière, M. L. 2002. A case of preferential bestiality (zoophilia). *Sexual Abuse: A Journal of Research and Treatment* 14: 83–88.

Gomes, C. M., Ribeiro-Filho, L., Giron, A. M., Mitre, A. I., Figueira, E. R. R. and Arap, S. 2000. Genital trauma due to animal bites. *Journal of Urology* 165: 80–83.

Good News Bible 1978. New York: American Bible Society.

Holden, T. E. and Sherline, D. M. 1973. Bestiality, with sensitization and anaphylactic reaction. *Obstetrics and Gynecology* 42: 138–140.

Hunter, J. A., Lexier, L. J., Goodwin, D. W., Browne, P. A. and Dennis, C. 1993. Psychosexual, attitudinal, and developmental characteristics of juvenile female sexual perpetrators in a residential treatment setting. *Journal of Child and Family Studies* 2: 317–326.

Itzin, C. 1998. Pornography and the organization of intra- and extrafamilial child sexual abuse. In *Out of Darkness: Contemporary Perspectives on Family Violence*, 58–79, ed. G. K. Kantor and J. L. Jasinski. Thousand Oaks, CA: Sage.

Jimenez-Jimenez, F. J., Sayed, Y., Garcia-Soldevilla, M. A. and Barcenilla, B. 2002. Possible zoophilia associated with dopaminergic therapy in Parkinson disease. *Annals of Pharmacotherapy* 36: 1178–1179.

Randall, M. B., Vance, R. P. and McCalmont, T. H. 1990. Xenolingual autoeroticism. *American Journal of Forensic Medicine and Pathology* 11: 89–92.

Rehan, N. 2003. Profile of men suffering from sexually transmitted infections in Pakistan. *Journal of Ayub Medical College* 15: 15–19.

Russell, M. D. 1996. *The Sparrow.* New York: Fawcett Books.

Sandnabba, N. K., Santtila, P., Nordling, N., Beetz, A. M. and Alison, L. 2002. Characteristics of a sample of sadomasochistically-oriented males with a recent history of sexual contact with animals. *Deviant Behavior: An Interdisciplinary Journal* 23: 511–529.

Sandnabba, N. K., Santtila, P., Wannäs, M. and Krook, K. 2003. Age and gender specific sexual behaviors in children. *Child Abuse and Neglect* 27: 579–605.

Serpell, J. A. 1996. *In the Company of Animals: A Study of Human–Animal Relationships.* New York: Cambridge University Press.

Sgroi, S. M. and Sargent, N. M. 1993. Impact and treatment issues for victims of childhood sexual abuse by female perpetrators. In *Female Sexual Abuse of Children,* 14–36, ed. M. Elliot. New York: Guilford Press.

Shenken, L. I. 1964. Some clinical and psychopathological aspects of bestiality. *Journal of Nervous and Mental Disease* 139: 137–142.

Shepard, P. 1996. *The Others: How Animals Made us Human.* Washington DC: Island Press.

Vollum, S., Buffington-Vollum, J. and Longmire, D. R. 2004. Moral disengagement and attitudes about violence toward animals. *Society & Animals* 12: 209–235.

Wang, P. 2004. The prosecution of Taiwan sexuality researcher and activist Josephine Ho. *Reproductive Health Matters* 12: 111–115.

Weigand, P., Schmidt, V. and Kleiber, M. 1999. German shepherd dog is suspected of sexually abusing a child. *International Journal of Legal Medicine* 112: 324–325.

Notes on contributors

Frank R. Ascione, Ph.D., is a Professor of Psychology at Utah State University and co-editor of two books on animal abuse. His most recent book is entitled *Children and Animals: Exploring the Roots of Kindness and Cruelty* (2005, published by Purdue University Press). He lectures nationally and internationally on topics related to animal maltreatment and its association with family violence and child abuse.

Andrea M. Beetz, Dipl.-Psych., Dr. phil., psychologist, conducts research on human–animal interactions, attachment and emotional intelligence. She studied psychology at the Friedrich-Alexander University of Erlangen-Nürnberg, Germany, where she also did her doctoral dissertation on *Love, Violence and Sexuality in Relationships between Humans and Animals.* She has also studied and conducted research on human–animal relationships in the USA (UC Davis, Utah State University) and the UK (University of Cambridge).

Gieri Bolliger, Ph.D., Attorney-at-Law, is a Research Associate of the Foundation for the Animal in the Law (Stiftung für das Tier im Recht, Zürich/Berne, Switzerland), guest lecturer at Zurich Law School on animal welfare law, and author and co-author of several books concerning Swiss and international animal welfare law.

Antoine F. Goetschel, Ph.D., Attorney-at-Law, is Director of the Foundation for the Animal in the Law (Stiftung für das Tier im Recht, Zürich/Berne, Switzerland), guest lecturer at Zurich Law School on animal welfare law, and author and co-author of many books concerning Swiss and international animal welfare law.

Hani Milestki, MSW, Ph.D., is a licensed, clinical social worker with a Master's degree in Social Work and a doctoral degree in Human Sexuality. She is a certified sex therapist (Diplomate) and supervisor by the American Association of Sex Educators, Counselors, and Therapists (AASECT). She served as the Director of the Human Sexuality Institute, at The Fogel Foundation in Washington, DC, has a private practice in Bethesda, Maryland and has been training and supervising other therapists and professionals in the field of sexuality.

Helen Munro, BVMS, MRCVS, graduated from the Faculty of Veterinary Medicine at Glasgow University in 1966 and specialized in Veterinary Pathology in 1974. She has lived and worked in Kenya, Hong

Kong, Fiji and also in the UK, where she was Head of Pathology for a large pharmaceutical company. She returned to academia in 1995 to pursue her studies of animal abuse, publishing in 2001, with her colleague Michael Thrusfield, a series of papers that provide the first description of the "Battered Pet."

Michael Thrusfield, MSc, BVMS, DTVM, MRCVS, Dip ECVPH, C Biol, FI Biol, graduated in Veterinary Medicine from the University of Glasgow in 1970. He worked in the East Indies, and then gained postgraduate qualifications in Tropical Veterinary Medicine at the University of Edinburgh, and in Virology at the Medical School of the University of Birmingham. After a period of time in general practice, he returned in 1976 to the Royal (Dick) School of Veterinary Studies, Edinburgh, where he was appointed epidemiologist. He was elected a Fellow of the Institute of Biology in 1996 for his contribution to veterinary epidemiology, and is a foundation Diplomate of the European College of Veterinary Public Health.

LaVergne, TN USA
23 October 2009
161814LV00001B/2/P

9 781847 883544